Professor Ian Lowe AO is president of the Australian Conservation Foundation, emeritus professor of science, technology and society at Griffith University in Brisbane, and adjunct professor at Sunshine Coast University and Flinders University. Professor Lowe has been a referee for the Inter-Governmental Panel on Climate Change, attended the Geneva and Kyoto conferences of the parties to the Framework Convention on Climate Change and was a member of the Australian delegation to the 1999 UNESCO World Conference on Science. He attended the UN convention in Copenhagen in December 2009. His previous books include *A Big Fix* and *Living in the Hothouse*.

A VOICE OF REASON

IAN LOWE

REFLECTIONS ON AUSTRALIA

UQP

First published 2010 by University of Queensland Press
PO Box 6042, St Lucia, Queensland 4067 Australia

www.uqp.com.au

Typeset by Post Pre-press Group, Brisbane
Printed in Australia by McPherson's Printing Group

National Library of Australia Cataloguing-in-Publication Data
Lowe, Ian, 1942–
A Voice of Reason / Ian Lowe.
ISBN: 9780702238475 (pbk)
9780702238222 (pdf)
Subjects: Environmental protection – Australia.
Environmental ethics – Australia.
Environmental policy – Australia.
Sustainable development – Australia.
Environmental policy – Australia.
363.70994

UQP uses papers that are natural, renewable and recyclable products made from wood grown in sustainable forests. The logging and manufacturing processes conform to the environmental regulations of the country of origin.

CONTENTS

INTRODUCTION

Even if you search for Caragabal in a road atlas, you may not find it. A speck on the map, it was a proverbial one-horse town when I was growing up there in the 1940s. It is in central western New South Wales – west of Grenfell, east of West Wyalong, south of Forbes – in the middle of bloody nowhere, a part of the world where it was said the crows flew backward to keep the dust out of their eyes. My mother was married to a baker who produced bread for the wheat-growing district. In the post-war 1940s, supplies were limited and commodities like tea, sugar and butter were rationed. Beer was also hard to get. As bakers could obtain yeast in those days, the men of the district would occasionally gather in our bathroom to produce (probably illegally) a tub of home brew to share. One of these meetings provides one of my earliest political memories, of which more later.

My other memories of Caragabal were of apocalyptic events, plagues of biblical proportions and regular dust storms, heralded by a red cloud in the western sky. Though everyone piled inside and closed doors and windows, a thin layer of fine red dust inevitably infiltrated the house. The plagues of grasshoppers and mice were extraordinary events, filling the air with a green mosaic in the former case and providing a moving grey carpet in the latter. I also remember one occasion when there was torrential rain, causing the local creek not just to flow but to teem with yabbies. The whole town came down to the bank of the now marvellously fruitful creek for the free food. These were rare distinctive events. Not much happened in Caragabal from one week to the next.

Each day, Monday to Saturday, one train passed through in each direction along the branch line from Forbes to Stockinbingal. I memorised the exotic names of the other stations along the line like Quandialla, Berrendebba and Pullabooka, all local Indigenous names even though there were no longer any signs of those people in that part of the country. The little rail motor that made half the trips wasn't very interesting, but I loved the 'mixed' train which went north on Monday, Wednesday and Friday mornings and south on Tuesday, Thursday and Saturday afternoons. It consisted of goods trucks and a dusty old passenger carriage. At Caragabal station, the engine left the carriage standing at the platform while it shunted the goods trucks, leaving supplies for the local area and collecting produce to be taken away. As a pre-school child I would sit at our gate and watch the operation, fascinated by the complex process required to link our little town with the world outside.

Travel to big towns like Grenfell and Forbes required significant amounts of the precious rationed petrol, so trips were rare and not lightly undertaken, usually the result of either a medical crisis or the need for long-overdue replacement of an essential household item. Caragabal was an isolated community, linked to the rest of Australia only by rail, mail and the miracle of radio. While there were a couple of local commercial stations, the ABC was our lifeline, as it still is for remote rural areas.

When I turned five, I went to the local two-room school. With other five-year-olds, I was taught my letters; we sat happily chanting 'a for apple, e for egg, i for ink, o for orange, u for umberella'. I learned to read much faster than the other Grade One kids and was soon transferred to Grade Two, then to Grade Three at the end of the year.

The local teachers told my parents that they had a very capable boy on their hands, but that was probably the least of their concerns at the time, with a business to run and my three younger siblings to manage. The school decided to hold me in Grade Three for a second year because they judged me too young to be put in with the big kids in the room for classes four to six. So I spent that third year of school mainly helping the younger children and voraciously reading the relief parcels of books which came to the school from the central library in Sydney.

Outside of school, the regular afternoon radio broadcast of the Argonauts was a joy. As Theophrastus 17 I sent off written pieces about everything from the process of bread-making to the plague of mice, collecting points toward the coveted award of Dragon's Tooth and the distant lure of the Golden Fleece. Sydney seemed a world away, just as remote from everyday life in Caragabal as the fabled journeys of Jason and his Argonauts in the Mediterranean of ancient times.

Caragabal was in many ways a metaphor for the Australia of the late 1940s. The country had its Anglo-Celtic attitude and suspicion of the Asian 'hordes' to the north reinforced by the narrow escape from invasion by Japan in the early years of the decade. Immigration was controlled by what was quite unselfconsciously called the White Australia Policy and some states were still removing children from Indigenous mothers in the hope of 'breeding out' Aboriginal people entirely. Interstate travel was difficult because one of the continuing legacies of our history as warring colonies was a variety of rail gauges, so travel between Sydney and Melbourne required changing trains at Albury. Overseas travel was almost unknown for anyone but prime ministers and the select few who represented Australia at cricket, rugby or Olympic events. Internal communications were slow, with the telegram the chief advance on waiting days for a letter, although a few homes and many businesses now had telephones. For people living in country towns, as a larger proportion of the population then did, travel to 'the big smoke' of the state capital was at most a once-a-year event to coincide with the Show: the Royal Show at Easter in Sydney, the Exhibition (or 'Ekka') in August in Brisbane, and so on. There was a distinctive Australian culture, in some ways a product of the early days of convicts and Irish settlers being reluctant to accept authority.

From the late 1940s to the mid 1950s, the US was gripped by anti-communist hysteria. Creative professionals were banned or imprisoned for their links to leftist organisations. After he was elected prime minister in 1949, Bob Menzies tried to institute a similar regime in Australia. He tried to ban the Communist Party and similar groups. The consensus in the home brewery bathroom of the Caragabal bakery was against the proposal. Our system was superior to communism, the men of the district argued, because we were free to make our own decisions. They recognised the inherent contradiction in Menzies's plan and the US practice: while claiming to defend 'freedom', they sought to curb freedom. A decision of the High Court, striking down Menzies's proposed law, and a bitterly contested referendum in which the Caragabal view prevailed, saved Australia from the worst of the Cold War divisiveness. I have often reflected on the common sense of those relatively uneducated people out in the wheat belt 60 years ago.

I began to notice significant change in Australia in the 1950s. The Snowy Mountains Scheme, a grand project to dam rivers and harness water that was running 'wastefully' into the sea, aimed to provide both hydro-electricity and irrigation water for inland areas. The scale of the scheme required bringing large numbers of migrants to Australia, mostly from non-English-speaking backgrounds. The aftermath of the war in Europe also brought many other migrants from Greece and Italy, in turn changing Australia's pattern of eating and drinking as coffee shops, wine bars and even such exotic places as pizza shops opened. While the main source of migrants to Australia continued to be the UK and Ireland, the new migrants were more distinctively different and did much to change the underlying culture. By the mid 1960s, Sydney and Melbourne had become more cosmopolitan cities; Brisbane and Perth were still like large country towns by comparison. Air travel had become less exotic and even university sporting teams occasionally took to the air for long trips to Tasmania or Western Australia.

Even so, when I left Australia in 1968 to study in England, we were still effectively an off-shoot of Europe. If we studied a foreign language at school, it was usually French; if a second language was considered, it was Latin or German. The national anthem was still 'God Save the Queen' and it was played even at the conclusion of the film at cinemas, causing patrons to pause halfway out of their seats to stand solemnly in respect of the British monarchy. Prime Minister Bob Menzies had seriously suggested calling the new currency unit the 'royal' when decimal currency was introduced in 1966; it was the first major impact of creeping Americanisation when public opinion forced a change to the 'dollar'. There was still intense suspicion of our Asian neighbours to the north, and the Australian Government had enthusiastically joined the US when they precipitated a civil war in Vietnam to 'halt the spread of communism'. Australia was considered so far away from the centres of thought that university staff were expected to spend sabbatical years in the UK or possibly the US, catching up with new developments. Almost anyone contemplating an academic career, as I was by the late 1960s, went overseas to do a higher degree.

As the old saying goes, the past is a different country where they do things differently. It is hard to imagine a greater contrast than the one I see between the Caragabal in which I grew up and the Copenhagen conference I have just attended. Where Caragabal symbolised the isolation and inward-looking world of Australia in the 1940s, Copenhagen symbolises the globalised world of the 21st century. Climate change is a global problem, caused by the combined action of the whole human population in using fossil fuels and clearing vegetation. It affects everyone to a greater or lesser extent: fuel use in Cairo and Copenhagen affects the climate in Caragabal; fuel use in Caragabal affects the climate in Copenhagen and Cairo.

As I prepared to face the cold and damp of Denmark in December, I read that the Australian Government contingent consisted of over 100 people. Additionally there were scientists, journalists and representatives of non-government organisations, both environmental groups trying to ensure progress is made and business representatives trying (mostly) to prevent agreement that would curb their profits.

Altogether about 200 Australians were in Copenhagen for the climate change conference; about five times as many as there would have been if each country's delegation represented its share of the global population.

Copenhagen is, of course, not the only sign of our integration into global systems. When US financial institutions collapsed because of irresponsible lending and world oil prices, our banks and other institutions were so threatened that the Australian Government felt obliged to provide guarantees. When a serious economic recession affected the Northern Hemisphere nations, our government sent each of us $900 to stimulate the local economy. Our cities are now socially and culturally diverse in ways that could only have been imagined when I was young. Even if we don't watch the world news on SBS and stick to the much more parochial ABC or commercial bulletins, the internet links us firmly into the global community. So we are all aware of such trivial events as a domestic dispute affecting the world's best golfer and the subsequent revelations about his very complicated personal affairs.

The crucial point is that we are now part of the global system, so a kid from Nambour was in Copenhagen as a 'friend of the chair', playing a central role at the invitation of the Prime Minister of Denmark in the attempt to craft an agreement, aided by his minister who was born in Sabah but educated in Adelaide, while a kid from Caragabal was also there as part of the scientific community, trying to keep the politicians honest and make them face up to their global responsibilities. It is a completely different world from the one in which I grew up.

There is another fundamental change in contemporary Australia. One day in Caragabal was much like another and one year was much like another; nothing seemed to change and nothing seemed likely to change. By contrast, Copenhagen exemplified the complex modern world in which change is inevitable. As various essays in this collection show, it has been clear for some time that the future cannot be a simple extrapolation of the past. Australia is now, whether we like it or not, whether we understand it or not, part of a global family. We face problems that are global in scope. There is no prospect, even in principle, of solutions that are not globally inclusive.

This book is organised into four sections, in each of which are gathered a selection of essays or columns I have written in the last 20 years. The section about science, technology and the environment outlines the basic issues of the processes of science, its reasons for claiming to be a special form of knowledge and the recent erosion of trust in science. The technology which allows most Australians to live much more comfortably than previous generations is based on the body of scientific knowledge, but deciding which science to apply and how to apply it involves economic factors, social values and political choices. The environmental problems we now face are a direct consequence of the growth in our population and our consumption patterns, so large-scale environmental issues like climate change are directly linked to our technological choices.

As I think it is clear that the economic and political issues are strongly and inextricably intertwined, they are the focus of the next section. Even though we have been warned for nearly 40 years that the scale and rate of growth would lead to serious problems, short-term politics is still driving choices that must lead to consequences we will not enjoy. Some recent economic trends are shown to be the product of politics and ideology rather than any sort of rational analysis. While the warnings of scientists have been largely ignored, the global financial crisis demanded the attention of decision makers. It has led, at least in some quarters, to a critical review of the standard assumptions of economics and politics.

I have devoted a section to culture and health because cultural factors determine which technologies will be adopted and how they will be used, what political choices are possible and what economic trends will be encouraged. Just as cultural factors determine how we live, they also shape how we die; broad patterns of community health as well as our average life expectancy are influenced by our diet, our pattern of exercise, social cohesion and even such intangibles as the levels of inequality. Much of my wellbeing and enjoyment of life is linked to such recreational activities as singing and playing sport, so I have included specific comments on those activities.

Finally, I argue that education is the key to our future. At the individual and societal level, education determines our capacity to respond to changes as well as our ability to shape the future. So how we educate young Australians (and, for that matter, how we enable Australian adults to continue learning) is not just vital to their personal futures, it is also crucial to our national future. In a world where the global economy is increasingly dominated by services and value-adding, the best indicator of a nation's future is the education level of its workforce.

As a postscript, I reflect on the prospects for developing a global response to serious environmental problems in the light of the 2009 Copenhagen climate change talks, generally seen as a disappointment.

In the last 20 years, I have written a lot of material: 12 books, 45 book chapters, more than 30 journal articles and over 600 columns for various publications. So I had a wide variety of pieces to consider for inclusion in this volume. I have tried to select writings which link together to tell a consistent story about the human condition in 21st-century Australia and the ways we can work to shape a better future. While scientists who write about environmental issues are sometimes accused of spreading doom and gloom, I am incorrigibly optimistic that awareness of problems will allow us to make wiser choices and create better futures.

Except where I have clearly indicated the presence of revisions by saying that an essay is 'based on' earlier writing, I have resisted the temptation to update material that was written some time ago. Many of the pieces in this collection were written and published as many as 20 years ago. Rather than revising them to make me appear more prescient than I was, I have only corrected changes introduced against my will by sub-editors and inserted a small number of historical notes where it seemed important to help a modern reader understand the 1990 argument. Putting this collection together was a depressing reminder of how little progress we have made since 1990 in tackling some of the big issues we face.

SCIENCE, TECHNOLOGY AND ENVIRONMENT

INTRODUCTION: SCIENCE, TECHNOLOGY AND VALUES

I grew up in a world which saw science as different, more reliable, a more trustworthy form of knowledge. At school I learned of scientific 'laws': not theories, but ironclad rules that could not be broken. My formal education in science was really an introduction to science as a body of knowledge. As a young professional I heard another scientist say in admiration of a colleague, 'When he measures something, it *stays* measured.' So I grew up seeing science as a body of permanent and reliable truth. But when I began doing scientific research, I soon realised the limitations of this view. The whole point of doing research is to extend knowledge, perhaps to question prevailing wisdom. So even if the existing science is totally solid, it must only be a partial and incomplete understanding of the world; if it were the last word, there would be no need for more research. Secondly, if some scientists are respected by their colleagues because of the quality of their work – 'it *stays* measured' – it implies that others are less impressive; there may be scientists who cut corners or who neglect to dot every i and cross every t before they publish their results.

There are also issues about what science we do. Social processes determine what research is funded from the public purse, while economic considerations influence the research and development in the private sector. Asking a question does not guarantee it will be answered, but not asking it makes it much less likely that an answer will be found!

We also live in a world where our material living standards have been dramatically improved by the application of new scientific knowledge in the form of technology. Our level of comfort is much greater than existed 50 years ago in this country, and much greater than exists now in many parts of the world. Technology that we take for granted gives us clean water, breathable air, a variety of safe food, the capacity to modify our living environment, opportunities for communication and transport. While we tend to take these services for granted, it was conscious political decisions that gave us sewerage systems to process our wastes, dams and treatment facilities to provide clean drinking water, public transport systems or roads or cycleways or footpaths, satellite communications or copper cables. Deciding which technologies to apply and how to apply them inevitably involves values-based choices.

More recently, it has become widely accepted that some areas of science are also affected by social processes. Scientific observations are always, to some extent, influenced by prevailing theories and therefore have subjective influences. We all tend to see what we expect to see. Our values can determine whether we see the glass as half-full or half-empty. There is much more likely to be disagreement in complex fields where we do not have all the data we need for a clear conclusion. So while competent scientists will all agree about a measurement that can be carried out under controlled conditions in a laboratory – the melting point of lead, or the electrical conductivity of copper – it is much harder to assess, for example, the impact of a new pressure on a complex system: whether the introduction of genetically modified crops threatens the food production system of rural Australia, or whether Australia would be better placed if we develop solar energy rather than nuclear power. Competent scientists can – and do – disagree about these sorts of issues.

More fundamentally, science is not a stable body of knowledge but is sometimes a succession of different worldviews. A new theory might be a refinement of the old, but sometimes it will be a radically different approach. When I was an undergraduate, 'continental drift' was populist superstition. The continents looked as if they could once have fitted together but the whole idea was rejected because scientists could not imagine continents moving about on the surface of the Earth. One crucial measurement, showing that the Atlantic sea floor was spreading and the continents were moving apart, launched the new theory of plate tectonics. Almost overnight, yesterday's superstition became today's respectable science. There have been a number of these scientific revolutions: the overthrow of the view that the Earth was the centre of the universe; the recognition of the great age of the Earth; quantum physics and the theory of relativity; the acceptance of the evolution of species rather than belief in creation by a divine being.

The acceptance or rejection of a scientific theory is not entirely a rational process, in the sense of being dictated entirely by the internal logic of the subject; values play an important part. Scientists are fallible humans. They are understandably reluctant to accept that their pet theory might be incomplete or invalid. This can be seen in the convoluted attempts by some to explain inconvenient new evidence. When I was a student, there was a vigorous debate between cosmologists about the relative merits of the 'big bang' theory and the alternative model of a 'steady state' universe. The data gradually reinforced the 'big bang', but some opponents devised more and more complex explanations to justify clinging to their alternative model of the world.

Perhaps the most famous example of values influencing perception is the reaction of Albert Einstein, one of the greatest physicists of all time, to the usual interpretation of quantum physics. We expect undergraduates to understand that there is a class of events at the sub-atomic level for which simple cause-and-effect relationships do not apply. It is only possible to make probabilistic statements, such as: 'There is a 50 per cent chance that this cesium atom will undergo a fission reaction in the next 13 days.' Einstein famously said that he could not accept 'that God plays dice', so he could not accept that the future could only be expressed as a statistical probability.

Some scientists are reluctant to accept that there is always some degree of subjectivity in science. They worry that it could lead to a sort of postmodern relativism in which any whacky theory is equally valid. I don't think this is a real fear: scientific theories stand or fall by their capacity to explain the results of observation or experiment. But in complex systems, there will often be more than one defensible interpretation of uncertain evidence. In the case of the science of climate change, in the 1980s there was uncertainty about whether it could be proven that the observed changes were being caused by human activity. By the mid 1990s, the body of research had forced most climate scientists to conclude that human activity, mainly burning fossil fuels and clearing vegetation, is directly responsible for climate change. This view was not universal, but it is certainly the conclusion of almost all climate scientists. The few who disagree have been supplemented in recent years by scientists from other fields, most often geology, and people with little or no scientific background.

I certainly don't suggest those who still deny the science of climate change can be ranked with Einstein; however, it is clear to see values driving their reluctance to accept the science. Some are unwilling to believe that humans can influence the complex natural systems of the planet. Others are very conservative people who are deeply hostile to the very idea that profitable enterprises might be curbed because of their environmental impacts. These deniers of the climate science are simply a special case of the general observation that science is always influenced by values. Scepticism, an honourable tradition within science, becomes denial when there is no longer an intellectually respectable case for disagreeing with mainstream science. Those who are in denial characteristically change their explanations as often as necessary to maintain their predetermined position. This seriously stretches the idea of what is respectable science.

Let me give a concrete example. If your local medical practitioner tells you that you might have cancer, it is obviously sensible to seek an opinion from a specialist in the field of oncology. Even if one expert confirms the diagnosis, you would probably seek a second opinion. It would be sensible to consult another specialist before consenting to surgery or other life-changing treatments. That is a legitimate sceptical response. If several of the best oncologists agree that you have a serious but operable cancer, it would be perverse to shop around the fringe medicine community until you found a quack who would tell you just to drink pomegranate juice and put your head out the window on a moonlit night – that would be denial. When almost all the best climate scientists in the world agree that human action is changing the global climate and threatening our future, it is just perverse to shop around for a retired geologist or an unsuccessful British politician who will assure you that the scientists have got it wrong and there is nothing to worry about. At the end of 2009, that was the stance of our Opposition Coalition in the national parliament.

As I was writing this introduction, there was a dramatic increase in the attention given by commercial media to climate science denial. Hackers accessing the email system of a UK university discovered that some scientists had been actively trying to prevent publication of dissident views, a practice contrary to the norms of science. It was also revealed that a small number of forward projections in the latest Intergovernmental Panel on Climate Change (IPCC) report were not solidly based on research that had undergone the usual peer review before publication. These were serious criticisms that demanded attention, but it is just ridiculous to claim they undermine the entire edifice of climate science. An analogy might be if I were to claim that Newtonian mechanics is invalid because I failed to sink a ball on a snooker table!

There is no serious doubt that greenhouse gases like carbon dioxide, methane and water vapour keep the Earth much warmer than it would otherwise be. Research published in the 19th century showed this and warned that we might increase average temperatures if we increased the concentration of those gases. In the 1980s, climate scientists first warned that we seemed to be producing those changes. Literally thousands of person-years of scientific effort went into examining and testing the hypothesis, as well as the alternative possibility that there was no link between human activities and the observed changes in the climate. By 1995, there was no serious doubt about the link: burning fossil fuels and clearing vegetation was changing the global climate. Worryingly, we are already seeing all the changes that climate models 20 years ago predicted we would see in the 2020s. There is increasing evidence that changes are feeding on themselves and leading to a risk of runaway warming. In that context, the attempts to muddy the water and generate doubt about the science of climate change border on the irresponsible, because they are eroding the political will to respond.

The pieces of writing I have assembled in this section are intended to show the social process of scientific inquiry, how science and technology can continue to improve our material living standards, how science has alerted us to the impact of our actions on the natural environment and how technology can give us a better future.

MEASUREMENT AND UNCERTAINTY: THE PROBLEM OF IONISING RADIATION

Based on a public lecture given in 1994

We are all exposed every day to small quantities of ionising radiation. The Earth under our feet emits radiation, much more if you live on granite than if you live in a sandy estuary. More comes from the building you live in and even the humans you encounter in your daily life. As well as this 'background' radiation, we are increasingly likely to be asked to undergo medical diagnostic procedures which involve ionising radiation. The risk from low doses is an important public health question. It concerns such issues as the acceptability of nuclear power, the safety of medical procedures and the desirability of permitting irradiation of food.

Clearly it would not be morally acceptable to do controlled experiments, irradiating different groups and observing their health changes over time. So we have to try to develop an understanding by analysing data from unintended exposure. As the body of information has increased, the permissible occupational exposure to radiation has been steadily reduced; in the US, which has a long record of regulating the acceptable dose, the maximum figure for today's workers is less than one-tenth of the dose allowed 80 years ago. The US criteria explicitly represent a value judgement, setting the unquantifiable health hazards against the equally imprecise benefits to society of having medical isotopes, nuclear power or nuclear weapons.

In the absence of hard data, there are two competing theories for the relationship between radiation dose and the risk of damage to the body. For high levels of exposure, there is a roughly linear relationship between radiation dose and damage suffered. One theory says that this can be extended down to low doses. Others point to the fact that there is greater exposure where towns are built in rocky areas, such as the New England plateau or the Granite Belt, than where people live in estuaries like the Brisbane valley. Since there is no hard evidence showing that it is riskier to live in Armidale than in Brisbane, there is an alternative theory which proposes a threshold value below which radiation does no harm.

As you might expect, people tend to choose the theory which suits their values. Most environmentalists would argue that there is no safe dose of radiation, whereas most proponents of nuclear power believe there is no evidence that the resulting low doses of radiation do any harm. At the extreme, this makes them quite happy to cook the books to get their desired result. In his book *Social Control of Technology*, British author David Collingridge quoted Lord Hinton, then head of the UK Atomic Energy Authority, openly admitting that the initial calculation of the possible radiation dose to nearby residents from a proposed reactor was based on an assumption that 1 per cent of the radiation would escape the containment vessel. When this approach found the doses to be unacceptable, the authority re-calculated with an assumption that 'more realistically' only *0.1* per cent of radiation would be released and 'established the fact that the site was perfect'! This sort of exercise leaves little room for arguing that the scientific assessment of risk was objective in any normal sense of the word: since the answer from the first calculation was unacceptable, it was effectively divided by ten to give a lower figure in order to get the site approved.

Since then, regulatory authorities have carried out detailed studies on the health of individuals as a function of the radiation exposure from their housing. Scientific opinion is now leaning very strongly toward the view that there is no absolutely safe dose of radiation, simply a steadily declining risk for lower doses. So the permissible levels of exposure represent an explicit value judgement, trading off the social benefits (real or alleged) of uranium mining or medical exposure against the small extra health risk. There is an obvious parallel in permissible blood-alcohol levels for drivers. The risk of an accident steadily increases with increasing alcohol use, with no detectable threshold. So the legal limit is a value judgement. Some societies set the limit at 0.08 milligrams per millilitre, some at 0.05, whereas some classes of drivers (taxi drivers, bus drivers, airline pilots) have a zero-tolerance rule.

The operating safety of nuclear reactors is a second example. If enough reactors are built and operated for long enough, there will eventually be good statistical evidence of the risk of serious accidents. Before the Three Mile Island core meltdown and the Chernobyl explosion, nuclear authorities produced improbable calculations to estimate, as one example, that the likelihood of anyone living near a reactor being killed by an accident was as low as one chance in 300 million. These claims were based on an assumption that the probability of catastrophic failure could be calculated by multiplying the risk factors for various events that would combine to cause the accident. The obvious problem is that serious accidents, almost by definition, happen as a result of an unforeseen combination of events.

The containment of radioactive waste is a similar problem. Given that the waste needs to be isolated for periods of hundreds of thousands of years, how could there possibly be convincing evidence that the management systems are reliable over those immense time scales? In the absence of hard data, the public were given assurances like that of the Australian Science and Technology Council, which said in a 1985 report that 'any return of radioactivity to the biosphere can be held to safe and acceptable levels over long periods (up to 1 million years) so that the maximum doses to the most exposed individuals would be a small fraction of natural background levels'. This is an assertion for which no evidence is available.

The health risk from exposure to a hazard is not a simple issue. It has been remarkably difficult to find hard evidence to prove a link between exposure to carcinogens and subsequent development of cancer. The case of tobacco is one where the evidence was first drawn together nearly 50 years ago. Most people now accept that exposure to tobacco smoke gives a dramatically increased risk of lung cancer, heart disease and many other health problems. But for decades the tobacco industry has affected not to recognise the risks and continued to promote their products aggressively, aided by a small group of decision makers. Senator Nick Minchin, who became notorious in 2009 for denying the science of climate change, was also responsible a decade earlier for a minority report to a Senate committee denying the link between smoking and health risks, a denial straight from the tobacco industry manual. The asbestos industry for years denied the link between occupational exposure and mesothelioma. Jonathan Braithwaite has documented a frightening number of instances of negligence or outright fraud in the testing of drugs, as well as instances of reluctance to withdraw drugs from the market when even alarming side effects became known.

Alvin Weinberg argued nearly 50 years ago that there is a special class of problems, such as the biological effects of low levels of radiation and the probability of catastrophic accidents in nuclear reactors, which he called 'trans-science'. They are, he said, questions of fact that can be asked in the language of science, but they cannot be answered by science because the data are simply not available. He said that scientists have a responsibility to understand when the questions they are asked cannot be answered with the same sort of confidence as we can measure constants in the laboratory. Failing to make clear the limits of scientific knowledge, Weinberg said, gives conjecture the appearance of scientific proof. Reviewing the 1985 Australian Science and Technology Council (ASTEC) report on the nuclear fuel cycle, Jane Ford concluded 'the same facts could easily have been used to reach a totally different conclusion'. Similarly, I published a *Quarterly Essay* in 2007 and criticised the more recent Switkowski report on a possible role for nuclear power in Australia. I think the report painted an unrealistically rosy picture for that technology. Where the complexity of the problem or the impossibility of conducting controlled experiments means knowledge is incomplete and gaps have to be filled by assumptions, there is no possibility of claiming that the science is objective. People with different values will use different assumptions and reach different conclusions. So an informed reader should always try to discern the values of the writer and take them into account to assess the validity of the conclusions.

THE 'HOLE' IN THE OZONE LAYER: A CASE STUDY OF COMPLEX SCIENCE

Based on a section of Living in the Greenhouse, *Scribe Books, 1989*

The story of the unravelling of the environmental effects of chlorofluorocarbons (CFCs) is a fascinating one, with a few ironic twists and some object lessons for the future. It began in 1970 when a British medical researcher, Dr James Lovelock, decided to take early retirement to give himself time to pursue other interests from his garage-cum-lab in the backyard of his house. It was, at least by British standards, in a quiet rural area. He immediately ran up against the entrenched snobbery of the British science establishment. No longer having a prestigious institutional address, he found it difficult to obtain research grants or even to have his scientific papers published. When he sought clarification of the rejection of one paper by a leading scientific journal, the editor told him in lofty tones that they were always getting crank papers from funny people who lived in the country.

No doubt that sort of objective peer review was behind the rejection of his 1971 application for a small research grant. He requested funds to build equipment that would measure the concentration of CFCs in the atmosphere. One referee said that the proposed apparatus could not be built and that, even if it could be, the proposed measurement was a complete waste of time. Having been spurned by the experts, Lovelock scrounged disused equipment and raided the family housekeeping money to buy odds and ends to build his equipment in the garage. Satisfied it was working, he arranged to travel on a ship supplying Antarctic scientists, and measured the atmospheric concentration of CFCs for a range of latitudes. The values he found were very small so, when he reported the results in a 1973 paper, Lovelock commented explicitly that the levels did not appear to pose any threat to the environment or to human health.

Fate then took a hand, as it so often does in the real world of science. On a rest day during a Vienna conference at which he presented his results, Lovelock went for a walk with a Dr Machta of the US space agency NASA, and a chemist from DuPont, the company that produced CFCs. After discussing Lovelock's data and asking the chemist about the production rates for the chemicals, Machta realised that the amount of CFCs still in the air was not much less than their total production up to that time. In other words, it seemed that almost all the CFCs that had ever been released were still in the air.

The second chance link was a conversation that took place a few weeks later during a coffee break at a conference in Florida. Dr Machta told Professor Sherwood Rowland, a chemist from the University of California, of his interesting calculation. Rowland was intrigued, and thought it would be worth trying to find out what happened to CFCs in the atmosphere. He wrote a grant application, seeking funds for such a study. It, also, was turned down. As luck would have it, Rowland had some uncommitted funds – enough to employ a post-doctoral fellow to work with him. He recruited Dr Mario Molina late in 1973 and presented him with a choice of possible projects. Dr Molina chose to work on the CFC issue.

Within a few months Dr Molina satisfied himself (and Professor Rowland) that CFCs are so stable that they do not react with any other material on the Earth's surface or in the lower atmosphere. Concluding that they would eventually be carried into the upper atmosphere, he realised that ultraviolet radiation from the sun could break down the compounds to release chlorine atoms. These could in turn produce a chain reaction, with each chlorine atom removing thousands of ozone molecules from the atmosphere. When he took into account the rate of production of CFCs, he estimated that between 20 and 40 per cent of the ozone layer could be destroyed.

To understand what could happen, you need to appreciate that the 'ozone layer' is simply a part of the upper atmosphere where the sunlight is intense enough to break up oxygen molecules (which consist of two oxygen atoms), allowing some of them to re-form as ozone (which is three atoms of oxygen). Ozone is unstable and breaks down naturally to oxygen. The 'layer' of ozone consists of the balance between ozone being formed by sunlight and its natural return to oxygen. The addition of chlorine to the 'layer' provided a new mechanism to break down the ozone, causing it to be lost much faster, and giving a different balance.

The analogy I find helpful is to imagine a sink with water running in from the tap and out through a plug-hole – the ozone being formed by sunlight is like the water running in, and the plug-hole is like the removal of ozone by its natural breaking down. The water level will be stable, representing a balance between water coming in and water going out. Now imagine a second plug-hole being opened, representing the chlorine atoms which can break down ozone. Water would flow out faster, so the level in the sink would fall; it would find a new balance between the constant inflow and the increased outflow. In just the same way, the new mechanism for removing ozone causes a new balance, with a lower level of ozone.

Molina and Rowland thought they must have made a serious

mistake. If the potential consequences of using CFCs were really as bad as they calculated, surely somebody would have already noticed, they thought. After an agonising re-appraisal trying to find the flaw in their logic, they published their work in June 1974 – only nine months after they began work on the problem, and only about a year after Lovelock's measurements were released. At once, Rowland found himself at the centre of media attention. He argued forcibly that the possible consequences of continuing to release CFCs were so serious that the products should be withdrawn from the market. DuPont, faced with public criticism of a profitable line of products, naturally disagreed. It found an ally in Lovelock, who regarded the controversy and the call for CFCs to be banned as typical overreaction by excitable American environmentalists. 'I think we need a bit of British caution,' he said. To illustrate his point, he recounted the story of US environmentalists blaming industry for measured high levels of mercury in tuna, until a more responsible scientist tested museum specimens that were a hundred years old and found them to contain similar mercury levels.

The story contains two exquisite ironies. One is that the same James Lovelock, who was calling for caution – making him briefly a folk hero among anti-environmentalists – later published influential research on what he called the GAIA hypothesis, the idea that the Earth is an organic, self-organising living system. This work made him a hero to environmentalists. On the other hand, the scientist who had tested the 19th-century fish and himself became a folk hero among anti-environmentalists – the more responsible scientist whose approach contrasted so favourably with the excitable Sherwood Rowland – was the *very same* Sherwood Rowland. In contrasting roles, he demonstrated the value to the community of having researchers who are not in anyone's pocket but will seek the truth and make it known, whatever discomfort that might produce for powerful interest groups.

The reason for the tradition of academic tenure in universities is to provide a guarantee that researchers will not run the risk of capricious dismissal if they antagonise powerful interest groups. As a former tenured academic, I had an obvious vested interest in defending the custom, but it is clear that the community benefits from having academics such as Rowland who are free to speak out on issues of importance. I certainly valued the freedom I had as a university scientist to contribute to the public discussion of important issues.

A similar argument applies to public funding of CSIRO, for many decades an independent scientific voice in Australia. Unfortunately, successive governments have steadily reduced the public funding of CSIRO and universities, forcing them to seek commercial support for their research. This, in turn, has made CSIRO and university scientists noticeably both less willing to speak up and less likely to be seen by the community as the impartial voice of scientific reason. This has been a significant loss. Some CSIRO managers have been so paranoid about the possibility of offending government ministers or agencies that they have effectively censored their scientists. Of course, there are powerful interest groups that benefit from the trend to a less-informed public debate, so we are unlikely to see a return to the earlier funding models that gave scientists more freedom.

When Rowland pursued his public attack on CFCs, DuPont challenged the theory with a corporate determination that ranks with the tobacco industry's denial of the health risks of smoking. They claimed there was no proof that CFCs would reach the upper atmosphere, that they would break down and release chlorine if they did reach the upper atmosphere, and that chlorine would remove ozone if it were released. They said it would be premature to restrict a profitable industry while research was still going on. Understandably, they were very coy about the level of proof that would convince them to abandon production of CFCs. Their advocacy of studied inaction was attacked by Rowland and others as the scientific community found no flaw in the theory.

Reviewing the debate in a 1978 book called *The Ozone War*, Lydia Dotto and Harold Schiff pointed out that doing nothing until the depletion of the ozone layer could be measured would inevitably mean that the situation would get much worse before it could be remedied. The reason for this is the long lifetime of the compounds in the atmosphere. They said it would be irresponsible to follow the course of a massive, uncontrolled experiment on the only atmosphere we have. Irresponsible it may have been, but that was exactly what we did. Only when the so-called 'hole' in the ozone layer was actually measured by Antarctic scientists in 1984 was the issue taken seriously.

Political pressure built, and in 1987 the first agreement to restrict the release of CFCs was signed. The Montreal Protocol was a crucial first step, but it did not solve the problem. In fact, it gave industrialised countries another decade to reduce their usage rates and allowed expansion in developing nations. The respected UK journal *New Scientist* said that the agreement was only signed because of industry and government fears that further research would reveal the depletion of the ozone layer to be much more serious, producing public demands for a more radical response. In fact, that happened, as the 'hole' in the ozone layer spread further north and reached southern parts of Australia. Ultraviolet intensities as much as five times the normal value were measured at ground level in Antarctica. Then the science showed that the ozone layer was thinning above Europe and North America, making the pressure for a stricter regime irresistible.

The original agreement was tightened at subsequent meetings in Stockholm and London. By the London meeting, even such a notorious exponent of market forces as British Prime Minister Margaret Thatcher was calling for action, telling delegates that they were not owners of the planet but tenants with a fully repairing lease! As Australia's then science minister Barry Jones commented, the CFC problem was a fundamental challenge to Thatcher's previous approach of trusting market forces and shouting at foreigners. Here was a problem that could only be solved by concerted international action, interfering in a market that was still demanding more of a profitable product. Moral pressure had no effect on the manufacturers, who were still selling and promoting their product 15 years after the science showed it to be posing serious risks. Despite this example by Thatcher, one of their own former stars, the Howard and Bush governments were still following a market-driven approach to the problem of global warming in 2007.

There are some broader lessons from the CFC story. The traditional peer-review approach to science did not support either the measurement of CFC levels or the modelling of its consequences, and the work was only done because the researchers involved had access to resources outside the normal granting mechanism. This shows the value to the community of a research capacity that does not depend on support from scientific elders. Second, the sequence of research activities that unravelled the problem was crucially dependent on informal contact between researchers in quite different fields of science. So we need to value those channels of communication. Science proceeds by researchers making discoveries and communicating them to others; until that process of communication takes place, the new knowledge has no value to the scientific community. Finally, the issue raises the fundamental question of the burden of proof.

In this case, it was assumed that the chemicals under suspicion were innocent until proven guilty. That is the approach we use in criminal law; if charged with an offence, I am presumed innocent and the case against me has to be proven beyond reasonable doubt in a court. The approach makes sense in the criminal code. Being locked up is such a serious penalty that it should only be imposed if there is no doubt about the guilt of the accused, especially as those seeking to prove guilt have the resources of the state at their disposal. It is a very different story when the question is whether a chemical is causing harm. In this case, those seeking to establish the 'guilt' of CFCs were individual scientists and voluntary groups, while those defending them from the charges were enormous companies and governments. This was also a very serious case, where CFCs were charged with the 'crime' of threatening community health and the balance of natural systems.

In hindsight, we have been far too generous to new chemicals such as CFCs. As we do with drugs for human medication, we probably should regard new chemicals as guilty until proven innocent. In other words, those who want to release new chemicals into the environment should have to show that they would do no harm. For the record, this might not have prevented the depletion of the ozone layer, since the science showing that it would be a problem had not been done when they were first released in the 1930s. As discussed earlier, Lovelock still believed in 1973 that the levels in the atmosphere were so low they would not cause any harm.

However, we now know that we were very lucky. As the Nobel prize-winning chemist Paul Crutzen has pointed out, it was simply a matter of chance that the chemicals developed and released were based on chlorine. They could equally well have been based on its sister element, bromine. Dr Crutzen has shown that the use of bromine compounds would have totally destroyed the ozone layer, with devastating consequences for life on Earth. We had a lucky escape from catastrophic changes. We should learn from this example; in future, we should be much more cautious about releasing millions of tonnes of new artificial chemicals into Earth's complex system.

We should also consider the fundamental legal principle involved. In criminal law, as discussed above, an individual is only convicted if guilt is proven beyond reasonable doubt. That is appropriate because the judgement affects the liberty of an individual or even, in some backward legal systems, their life. In civil law, by contrast, cases are decided on the balance of probabilities. I believe this is far too generous to polluters, now that we know that the lives of individuals and the survival of entire communities can be endangered by the release of hazardous substances. Those who want to release new substances should not only bear the burden of proof, but should have to prove beyond reasonable doubt that their proposal will be safe. The expert witnesses called to testify should also be held accountable if their advice turns out to have negligently exposed others to risk.

Even then, we should still reserve the right to review decisions in the light of new evidence. After all, CFCs seemed harmless when they were first developed in the 1930s. When scientific research showed that they would cause serious environmental damage and risk human health, the manufacturer was able to stall action by appeals to pragmatism and calls for further research. In retrospect, the delay was unconscionable, and it resulted in needless damage to both human health and natural systems. We should learn from this lesson of history. The ozone layer is now beginning to repair itself, and is likely to be back to normal by about the middle of this century. But the CFCs will remain in the atmosphere and continue to contribute significantly to climate change for decades at least, possibly centuries.

SCIENCE, TRUST AND THE ALLEGEDLY IGNORANT PUBLIC

Edited extract from Quarterly Essay 27, Reaction Time, *Black Inc., 2007*

Many scientists and engineers subscribe to the 'deficit model' of public understanding. It essentially says that the public are hostile to advanced technologies like nuclear power and genetic engineering because they are uninformed. Qualified scientists and engineers understand these complex questions and know that the benefits exceed the risks. The public should trust the experts who know better. If the public is irrationally unwilling to trust the experts, effort should be put into community education to dispel the ignorance and produce a new state of informed bliss in which all will cheerfully accept the advanced technologies.

Unfortunately, there is a particular problem that could be called the May Effect. Dr Bob May, now Lord May, former British Chief Scientist and now President of the Royal Society, gave a fascinating paper to the UNESCO World Conference on Science in 1999 looking specifically at the question of attitudes to genetic modification of food. He reminded those attending that most scientists believe in the deficit model, so they think that those who are informed understand the benefits and welcome the new technology while the ignorant are needlessly worried. He then looked at international comparisons of attitudes. The highest level of support for genetically modified food was in the US, where the level of scientific literacy is the lowest in the Organisation for Economic Cooperation and Development (OECD), with about a quarter of all adults believing that the Earth and all species were created in six days only a few thousand years ago and a significant number still thinking that the Sun goes round the Earth! As you move up the scale of increasing scientific literacy, May observed, you get decreasing support for genetic modification, with the opposition strongest in the Scandinavian nations which have the world's highest levels of scientific literacy.

His explanation for this unexpected effect was that only the ignorant believe that scientific advance is an unmixed blessing. If you understand the science, he said, you will be aware that it is always some sort of Faustian bargain; there are always costs as well as benefits. So, he said, we should not be surprised that those communities which have the highest levels of scientific education are the least likely to believe the bland assurances that all will be well.

The practical example of citizens' juries illustrates the May effect. The Australian Consumers' Association (ACA) conducted an exercise concerning genetic modification of food. A panel of citizens, screened to ensure they did not come with strong preconceptions one way or the other, went through a very intensive learning exercise. The panel chose experts with different views to deliver briefings and answer questions. At the end of the massive learning exercise, the members of the panel were much better educated about the issue than we could ever expect a random sample of the public to be – and there was no significant change in their attitudes toward the technology. In broad terms, those who began the exercise not well informed but generally sympathetic to the technology emerged at the end very well informed and sympathetic. Those who began the study not well informed but generally hostile reached the end-point very well informed and hostile. That reinforces the finding that our support for complex technologies, or our hostility toward them, is mostly based on our values rather than our detailed understanding (or lack of it).

This cautionary tale should be a warning to all those who suppose that public hostility to genetic engineering or nuclear power will be dissolved by education. Some of the most trenchant opponents of the nuclear push have doctoral qualifications in physics and could not remotely be accused of failing to understand the issues. Scientists such as Dr Alan Roberts, Dr Mark Diesendorf, Professor Jim Falk, Dr Jim Green and myself fall into this category; we all have strong backgrounds in physics and know how risky is the course being advocated. We also understand the science of climate change and the urgency of a response; in my case, I have been advocating reduction of our greenhouse pollution for 20 years. But we see nuclear power as a foolish response to the problem.

SILENCING DISSENT IN THE RESEARCH COMMUNITY

Adapted from a chapter in Silencing Dissent, *Clive Hamilton and Sarah Maddison (eds), Allen & Unwin, Sydney, 2007*

The 1999 World Conference on Science in Budapest was the first UN conference on science and its applications for 20 years. Jointly organised by UNESCO and the umbrella body for national academies and learned societies, the International Council for Science (ICSU), it called for a new social contract between science and society, with a formal code of ethics for scientists along similar lines to the Hippocratic oath taken by medical graduates. Nobel laureate Joseph Rotblat called for a code of ethics; he wanted to go much further than obvious basic principles like honesty and respect for human dignity. He argued that scientists should respect the global environment and take into account our responsibility for future generations. A paper by ICSU endorsed the idea of international guidelines.

Science in Australia seems to be moving in the opposite direction. Within CSIRO, there is now a culture of managerialism so wary of offending government that scientists have been instructed not to comment on issues that have policy implications. Even within universities, once prized for their belief in academic freedom and their willingness to protect academics from victimisation, there is now increasing pressure to conform. The policy line is often set on the basis of ideology or whim rather than detailed analysis. Science is urged to get on board the policy train. Those who support it cheerfully speak out, but those who know it to be wrong are intimidated into silence. Those who know do not speak, while those who speak do not know – or, even worse, pretend not to know.

A striking example was widely publicised by the ABC series *Four Corners* in February 2006. Professor Graeme Pearman is one of Australia's most distinguished atmospheric scientists. He has been involved in the measurement of greenhouse gas concentrations and the modelling of climate change for nearly 30 years, and as a result he has an outstanding international reputation in the field. Within Australia, he has been recognised by election as a Fellow of both the Academy of Science and the Academy of Technological Sciences and Engineering. The CSIRO Division then headed by Pearman organised two landmark conferences in the 1980s: Greenhouse '87, a major scientific conference to explore the implications of the scale and rate of climate change being predicted by CSIRO models, and Greenhouse '88, a national conference aimed at a broader audience to increase public awareness of the problem. He has continued to communicate the science to politicians and public alike, including briefings for many government agencies.

In 2004 we were both involved in a scientific initiative to draw together the disturbing evidence that climate change is already having serious impacts on Australia, with much worse to come if we continue to neglect the issue. The report of the Australian Climate Group, *Climate Change Solutions for Australia*, looked at environmental, economic and social impacts, including effects on community health. It covered the evidence that climate change is already happening, the concerns about future developments and some suggested responses. In retrospect, these were hardly radical suggestions. By 2005, the Australian Medical Association was joining the Australian Conservation Foundation to make similar suggestions, based on analysis of the alarming consequences for public health of the government's inaction. By early 2006, the CEOs of six large companies including BP, Westpac and Origin Energy were prepared to call publicly for the same sorts of measures: a long-term aspirational target of reducing greenhouse gas emissions by 60 per cent by 2050, a short-term binding target and market mechanisms to provide an incentive for business to invest in the new technologies needed.

As we prepared for the release of the report in mid 2004, we developed a media strategy to maximise its impact, with a central role for Pearman as a leading authority. At that point, he was pressured by those above him in the CSIRO hierarchy to confine his public remarks to the climate science and refrain from commenting on the policy issues. This approach was defended by Pearman's immediate CSIRO line manager, Dr Steve Morton, on *Four Corners* when he said the organisation's policy is that scientists can comment on science but not on policy. This approach was seen in action when another CSIRO scientist, Dr Kevin Hennessy, was asked questions about climate change. He said that global warming is a serious problem, that it could have disastrous effects and that we know what is causing the problem, but he was then placed by the CSIRO policy in the ridiculous situation of being unable to comment on which responses would be the most effective!

The message to Pearman from the CSIRO management was very clear: do your science, publish in learned journals, answer questions from the media about the technical issues of climate modelling, but do not say anything that could be interpreted by the most hypersensitive minister as implicitly critical of government policy. As Pearman puts it, 'As a climate scientist, I might inform [media] that the lifetime of carbon dioxide in the atmosphere means that the only way of stabilising global climate is by reducing emissions by 50 per cent by 2050 and by 80 per cent by 2100. In the current environment, that is seen as commenting on the government policy of not setting reduction targets.'

CSIRO management were apparently unhappy with Pearman's

public statements as part of the launch of the climate group's report, even though the report made no specific reference to the Commonwealth Government and Pearman stuck scrupulously to the limited role specified. He was subsequently made redundant, along with some other senior climate scientists in the division he had led. I know of ten leading scientists beside Pearman who have left the division in the last few years. Pearman says it was made very clear that management did not want him around, even banning him from entering the building in which he had worked for decades!

We now know that this is part of a broader pattern of the Howard Government making extensive efforts to control public information about the effects of climate change. The *Four Corners* program revealed that CSIRO management, under pressure from the government, had attempted to gag its scientists from speaking publicly about their research on climate change. Pearman said he was censored perhaps half-a-dozen times in the year before he was forced out of his position.

University scientists engaged in research into renewable energy are also intimidated. According to Philip Jennings, professor of energy studies at Murdoch University, renewables researchers believe they will lose their research funding if they are seen to criticise government policies on climate change and energy. So researchers are effectively intimidated into self-censorship, the most insidious form of censorship because it is totally invisible.

The Cooperative Research Centres (CRC) program was established by Professor Ralph Slatyer when he was the government Chief Scientist in the late 1980s. It was an imaginative initiative to address a long-standing problem in Australia: the repeated failure to turn our excellent science into practical innovations. Slatyer argued that all the incentives in universities and government laboratories encouraged pure science, leading to publishing in scholarly journals. Working with partners to turn research results into practical applications, either commercial ventures or developments in the public interest, is difficult and often unrewarded. His proposal was based on the adage that trying to direct academic researchers is like trying to herd cats – but you can make cats go where you want with a bowl of cream. The CRC scheme provided attractive research funding for proposals that brought together scientists with organisations that could turn the research results into practical applications – corporations, industry groups or public agencies.

In the early rounds, there was a healthy balance between commercial projects and those aimed at the public interest. I was a member of the committee when it funded a Centre for Renewable Energy. Other significant public-interest centres worked on integrated pest management, tropical rainforests, the Great Barrier Reef and the coastal zone. After the election of the Howard Government, every one of those centres saw its funding end. The CRC program moved decisively toward commercial outcomes. The changes are significant because the perception of what constitutes the public interest is intensely political. So funding research in the public interest leads inevitably to debate about where that interest lies. By steering research funds away from activities promoting the public interest, the Howard Government replaced the concept of the public interest with an economistic view, implicitly equating the good of the private sector with the good of the community. Of course, the change also steered research funds away from those dedicated individuals who were working for the public good rather than commercial gain, effectively muzzling them from participating in public debate by reducing the chance they had new results to advance.

Given the huge problems we face, we need to encourage new ideas and support challenges to conventional wisdom, not suppress them. Most senior people I know in industry and commerce support that view. Dr Graeme Pearman quoted one as saying in reference to the research community: 'What industry wants is independent and fearless advice.' So would a government concerned about our future. I have been forced by the scientific evidence to conclude that the current approach has quite fundamental problems, so I see the silencing of dissent as a national tragedy. The short-sighted policy systematically deprives our society of the innovations and new knowledge we need to avert our continuing decline.

THE MISMATCHED DUO OF SCIENCE AND STATISTICS

The Weekend Australian 31 March–1 April 1990

Science has a reputation that gives it a special status. The process of producing scientific knowledge is seen to give it a measure of authority and permanence. This issue of the permanence of scientific knowledge has been raised by the recent analysis of a cluster of leukaemia cases in England that occurred near the nuclear facility at Sellafield. The study found that workers exposed to low levels of radiation have six or seven times the expected probability of having children who contract leukaemia. This raised two important issues: the acceptability of statistical evidence and the validity of radiation exposure standards.

The acceptance of evidence which is mainly statistical has been an important issue in other public health debates, such as the risks of smoking tobacco. There has been evidence for decades now that tobacco is linked to lung cancer. Smokers also are noticeably more likely to suffer from a variety of other health problems, ranging from bronchitis to heart disease. The statistics show a convincing relationship between the level of exposure and the health risk. Thus those who smoke fewer than 15 cigarettes a day are about eight times as likely to get lung cancer as non-smokers, while those who smoke more than 25 cigarettes a day are 25 times as likely as non-smokers to get lung cancer.

There are two problems with statistical evidence. The first is that it can run counter to personal experience. If you personally know ten smokers, none of whom has lung cancer, you might conclude that smoking is an acceptable risk. You are very unlikely to know personally a sample large enough to be representative. It is human nature to be affected more by people we know than by statistics of the whole population. Even if *CHOICE* surveys show a given model of car to be the most reliable, you are likely to be wary of it if your neighbour has one that is always off the road for expensive repairs.

It is also true that association between two effects does not prove that one is causing the other. Anthony Tucker wrote 20 years ago of the statistician who concluded after a survey that attending London strip shows made men go bald! It is, apparently, true that the men attending London strip shows are more likely to be bald than a random sample of English males. It is clearly not valid to conclude that the taste in entertainment of these men has affected their hair. It would be equally absurd to conclude that the sight of large numbers of balding men caused women to take their clothes off. Clearly there is an association between what is happening on the stage and the composition of the audience, but the link does not prove that the performances cause baldness.

The point of this analogy is that there are still some who argue that the statistics alone do not prove that smoking causes lung cancer. The tobacco industry says that the link is purely circumstantial. This is an understandable policy, given the commercial need to find replacement customers for the smokers who quit, not to mention the estimated 17,000 who die each year in Australia from smoking-related causes. The National Health and Medical Research Council concluded in 1986 that smoking in shared spaces should be banned because of the health risk to non-smokers of inhaling the smoke of others; the tobacco industry does not agree that unacceptable health risks are revealed by the data on passive smoking.

There is a link between the health risks of smoking and the case of the Sellafield workers. Both pose very difficult problems for scientific inquiry. In the case of smoking tobacco, there is now a depressing accumulation of data to show that it is associated with poor health. While it is still difficult to isolate the harmful substances in tobacco smoke, it would certainly not be morally acceptable to conduct controlled experiments in which some humans were exposed to suspect chemicals to see if they got lung cancer. The same problem applies to ionising radiation. There is ample evidence of the unfortunate effects of large doses of radiation. At the other extreme, everyone is constantly exposed to low levels of radiation from a variety of sources: the Earth; buildings we use; luminous watches; even other humans. The issue of the health risks of radiation was raised in an acute form by the Chernobyl accident. As the cloud of radioactive debris spread across Europe, scientists were inevitably asked what the health risks were. Estimates of the eventual number of extra cancer deaths ranged from 5000 to 500,000.

Why can't science give us a better estimate? Again, it is obviously not morally acceptable to conduct a controlled experiment in which different groups are irradiated by different amounts so that differences in their consequent suffering can be measured. The relationship between radiation and its effects can only be studied indirectly or as a result of unintended exposure. As a result, occupational standards for exposure to radiation are a value judgement, involving a balance between health risks that cannot be accurately measured and the equally unquantifiable benefits of the nuclear industry. It is therefore a judgement on which different scientists can – and do – legitimately disagree.

It is also a judgement which has changed over time. Levels of exposure which once were tolerated are no longer acceptable. The Sellafield case has raised new concerns, because workers in nuclear facilities are monitored carefully. The levels of radiation to which these workers had been exposed was known, and was below the level regarded as perfectly safe. The evidence that their children are disproportionately likely to contract leukaemia suggests that there are risks associated with levels of radiation previously seen as harmless.

This does not represent a failure of science, but it does show clearly that scientific knowledge is not permanent. Further research is always likely to refine our understanding and reveal previously unsuspected risks. This is a solid argument for erring on the side of caution in exposing workers or members of the public to new health hazards.

BOFFINS DESERVE A SPORTING CHANCE

The Weekend Australian, 11–12 May 1990

Most Australians admire sporting heroes who have succeeded on the world stage, especially if they have struggled against great odds. Swimmers, runners, golfers and tennis players who have won at the highest level are household names. World champions are honoured even in such minority perversions as motor sports or racing overpriced yachts. Many still forgive Alan Bond's bizarre business dealings because he was largely responsible for prising the America's Cup loose from the bony grip of the New York Yacht Club.

Scientific stories attract much less attention. Few Australians know that we have won more Nobel prizes than Japan. Fewer still can name even one of the Aussies who have achieved this highest scientific accolade. Olympic gold medals are far more common than Nobel prizes, yet attract much more attention. This is partly because sporting events catch the public imagination. As a cartoon character once said: how much would you pay to watch a scientist?

We also have a deep-seated anti-intellectual attitude which is suspicious of any achievement in mental pursuits. Very few Australians are aware of the success of Monash University mathematician Dr Andrew Prentice. *The Australian*'s Higher Education Supplement recently gave prominence to a report on his work, but generally it has been ignored in the popular media. It deserves to be recorded in the annals of great struggles against the odds, in the same league as the triumphs over the New York Yacht Club, Heather Mackay's long domination of the world of squash or Robert de Castella's stirring marathon victory in Brisbane at the 1982 Commonwealth Games.

The most recent edition of the Australian science journal *Search* contains an article by Dr Prentice about his work. It is interesting science, but it is also a fascinating tale of his struggle against the entrenched power of the scientific establishment. Dr Prentice specialises in modelling planets and their moons using a modern version of a theory first put forward by the French mathematician Marquis de Laplace almost 200 years ago. He believes that the planets in our solar system have condensed from rings of gaseous material thrown out from the Sun when it was much younger and really just a spinning cloud of gas. The modern version of the theory suggests that the gas rings were formed by turbulence in the gas cloud which was our young Sun. Some recent observations of very young stars seem to bear out this idea, but the theory has not yet been accepted by the senior figures in that field of science: the old men of the tribe.

For years – because of this disagreement – Dr Prentice has found it difficult to get his work into the scientific literature. When a scientific paper is sent to a journal, the standard process of review involves the paper being sent for comment to referees who are regarded as eminent in the field. In this case, the great and the good were implacably hostile, and Dr Prentice's work was repeatedly rejected. The editor of one leading scientific journal even went to the trouble of writing a full-page editorial defending his refusal to publish articles submitted by Dr Prentice.

When the Voyager space probes were launched, their mission was to collect data from the outer planets. This gave Dr Prentice a golden opportunity to test his theories. He calculated the characteristics he expected of the moons of two outer planets, Uranus and Neptune. He had great difficulty getting his predictions into the scientific literature. After several rejections, he finally got the work published shortly before the Voyager probes reached the remote parts of our solar system.

The results were a stunning vindication of Dr Prentice's work. For example, he had figured out that there would be two moons, or groups of moons, orbiting the planet Uranus: one about 68,000 kilometres from the planet and the other about 89,000 kilometres away. In December 1985, Voyager 2 identified a satellite 66,000 kilometres from Uranus. Two weeks later, it discovered a belt of small moonlets; the average distance from the planet was 87,000 kilometres.

In 1989, the Voyager program yielded information about Neptune's moons. Dr Prentice had predicted that there would be a family of satellites – possibly as many as eight – 55,000 kilometres to 125,000 kilometres from the planet. The space probe found six moons between 50,000 and 118,000 kilometres from Neptune, with the exact orbits of these satellites agreeing closely with those calculated in advance. The theory also made correct predictions about the properties of these moons.

As well as explaining observations, a useful theory should also make predictions which can be tested. It is argued that a theory can only be regarded as scientific if it makes predictions which can be tested against the observable world. Dr Prentice's story, his struggle against the odds to show the validity of his theory, is the sort of stuff of which legends are made. It also has an important moral in these days of increasing centralised control over university research activities. We must maintain a capacity for research which challenges accepted orthodoxy. By the normal standards of his peers, Dr Prentice's theory was seen as not worth supporting.

Universities are steadily losing their capacity to fund research which does not attract support from granting bodies. As a result, unconventional ideas and theories are less likely to be explored. The increasing dependence on central funding bodies – however competent and well-meaning – carries with it the danger of an orthodoxy which precludes radical challenges and rules out the possibility of major leaps forward.

Not all unorthodox work is inspiring, but some is. The story of Dr Prentice shows that we should support some of those brave individuals who struggle against the tide of opinion.

SUSTAINABILITY SCIENCE

ABC Radio, Ockham's Razor, *June 2001; published in* Asimov's Elephant, *Robyn Williams (ed), ABC Books, 2003*

Science needs a fundamentally different approach if we are to achieve the goal of sustainability. That was the conclusion of a recent workshop at Friibergh in Sweden. It brought together two dozen scientists from all parts of the world, natural and social scientists, old and young researchers, men and women, united by their concern about present trends. I was among the invited participants.

We all accepted the conclusion of the recent UNEP report, *GEO2000*, which said that the world's present development path is not sustainable. We are trying to meet the needs of a growing population in a globalising, unequal and human-dominated world. Our efforts are exerting unsustainable pressures on the Earth's essential life-support systems. The complexity of the challenges ahead is clearly evident in the worrying interactions of climate change, loss of biological diversity, increasing poverty and growing inequality. Meeting fundamental human needs while preserving the life support systems of Earth will require significant changes. The human population is likely to increase by 50 per cent or more before it stabilises. We can also expect per capita demand for goods and services to continue to increase. There are still about 1 billion people without clean drinking water and nearly 3 billion without adequate sanitation, while about half the people alive have never made a telephone call or ridden in a car. So future demand for goods and services for the human population could well be double the present level, which is already stretching the capacity of natural systems. The Friibergh workshop concluded that the challenge requires development and promotion of a new field which is being called 'sustainability science'.

Sustainability science recognises that our understanding of nature–society interactions is still limited. There have been substantial advances in recent decades through work in the environmental sciences that factors in human impacts, and work in social and development studies that takes account of environmental influences. But we still have to accept that modern science can be described as islands of understanding in oceans of ignorance. We are constantly engaged in land reclamation, but there is no chance of filling in the oceans. So we need to set some broad priorities for our limited scientific effort. At the top of the list should be the urgent need for a better general understanding of the complex dynamic interactions between society and nature. That will require major advances in our ability to analyse the behaviour of complex self-organising systems, as well as developing better understanding of the irreversible impacts of interacting stresses. We need to work at multiple scales of organisation and consider the impacts on natural systems of various social actors with different agendas, ranging from environmentalists standing in front of trees to bulldozer drivers pushing them over and Cabinet ministers telling us it is justified because it promotes economic growth.

Case studies from all the inhabited continents show that many of our serious environmental problems are the direct result of applying narrow specialised knowledge to complex systems. Agronomists have advised farmers on fertiliser use to improve pasture, but the changes have put unacceptable nutrient loads on waterways. Expert advice has allowed fishing vessels to catch more seafood, leading to depletion of fisheries. Irrigation systems have made it possible to grow new crops, but have also deprived streams of the flows needed to maintain riverine ecologies. Species introduced to control one pest have driven other native biota to extinction. Coal-fired power stations provide cheap electricity, but their carbon dioxide emissions are now changing the global climate.

These conclusions came as no surprise to me, because the 1996 Australian State of the Environment report made a similar point about our serious problems. They are the combined effect of various pressures: population growth and distribution; lifestyle choices; the technologies we use and the demands they make on natural systems. While it is obvious to anyone who thinks about it that we are very unlikely to achieve our goal of sustainability if the population keeps growing, economists and politicians continue the mindless pursuit of growth as if it were an intrinsic benefit – but that's another story. The State of the Environment report also said that complex environmental problems have only been successfully solved where there was a systematic approach, combining all facets of the complex issue. Failures were usually piecemeal efforts that focused on one aspect of the problem to the exclusion of other, equally important, aspects. And the serious problems identified at Friibergh as the consequence of applying narrow specialised knowledge are all present in Australia: nutrient loads in waterways as a result of the application of fertiliser; depletion of fisheries because of modern technology in the fishing industry; river systems in crisis because of overzealous extraction for irrigation; introduced species out of control and causing chaos – and, of course, large scale salinity and biodiversity loss as a result of land clearing. So there can be no doubt about the worrying conclusion that great damage can be done by the application of narrow, specialised science without an appreciation of the complexity of natural systems.

We need to address these issues through integrated scientific efforts that focus on the social and ecological characteristics of particular places or regions. The Friibergh workshop formulated an initial set of core questions for sustainability science. These address the fundamental character of nature–society interactions, our ability to guide those interactions along more sustainable trajectories, and ways to promote the social learning needed to navigate a transition to sustainability.

By structure and by content, sustainability science must differ fundamentally from most science as we know it. The traditional scientific method was based on essentially sequential phases of scientific inquiry such as conceptualising the problem, collecting data, developing theories and applying the results. But these familiar forms of developing and testing hypotheses have run into difficulties as we study complex non-linear systems with long time lags between actions and their consequences. And the problems are complicated by our inability to stand outside the nature–society system. So we cannot even in principle be objective observers of the system. Think of the parallel of two people with different allegiances who watched a football match. If you talk to them, it can be difficult to believe they were watching the same game. They will differ about matters of fact, such as whether the ball was over the line, about matters of judgement, such as whether a foul was committed, and sometimes even about the visual acuity of the official in charge – or the official's integrity, or even the marital status of the official's parents! But our situation is worse than those biased observers, because we are actually out on the field with our faces in the mud! We can't even see the whole game and our inevitable interest in the outcome affects the way we see what is going on around us. As part of the nature–society system, we cannot even in principle stand outside it and be an objective observer.

So we have to accept that our engagement with complex natural systems cannot be based on the old model for rational objective science. The traditional sequential steps must become parallel functions of social learning, additionally incorporating the elements of action, adaptive management and policy as experiment. Sustainability science therefore needs to employ new methods such as semi-quantitative modelling of qualitative data and case studies, or inverse approaches that work backward from undesirable consequences to identify pathways that avoid those outcomes. Scientists and practitioners need to work together to produce trustworthy knowledge that combines scientific excellence with social relevance.

Meeting the challenge of sustainability science will also require new styles of institutional organisation to foster interdisciplinary research and to support it over the long term, to build capacity for that research and to integrate it into coherent systems of research planning, assessment and decision support. The present approach is too haphazard and piecemeal. It is chastening to remember that the two crucial pieces of research that identified the problem of ozone depletion both failed to attract support through the conventional peer-review process. That science was only done because the researchers concerned – James Lovelock and Sherwood Rowland – had access to discretionary funds. We can't afford to leave such important issues to chance – especially in a world where fewer and fewer scientists have discretionary funding that allows them to follow where the science leads. We need a better process to involve scientists and practitioners in setting priorities, creating new knowledge, and testing it in action.

This will require systems to integrate work situated in particular places and grounded in particular cultural traditions with broader networks of research and monitoring. Examples discussed at the workshop included mountain development in the Himalayas, global El Nino-Southern Oscillation (ENSO) forecasting and decision support systems in Africa, scientific support for the European convention on acid rain management and the Yaqui Valley study of land-use change in Mexico. I don't have any doubt you can think of local examples. As one obvious example, the independent review of the Natural Heritage Trust program was devastating in its criticism of the way the resources were frittered away without any strategic focus or scientific setting of priorities. Like the recent road funding fiasco, the Heritage Trust money was simply spread around electorates the government wanted to retain at the next election. The restoration of our precious natural resources deserves better than an opportunistic pork-barrel approach – but that's another story.

The emerging field of sustainability science will need to move forward along several pathways. There will be discussion within scientific communities, north and south, of the approach, its key questions and institutional needs. We will need to reconnect science to the ongoing political efforts for sustainable development, using in particular the forthcoming Rio + 10 conference that will review developments since the 1992 United Nations Conference on Environment and Development (UNCED). And who knows, we may eventually get politicians to accept the science and give us a functioning treaty on global warming, despite the concerted efforts by the US Government and ours to sabotage the treaty for the sake of short-term commercial interests.

Around the world, in large and small groups, research is underway and accelerating on the core questions I mentioned earlier: the fundamental character of nature–society interactions; our ability to guide those interactions along more sustainable trajectories; and ways to promote the social learning needed to navigate a transition to sustainability.

We have to develop urgently the mechanisms to nurture those research activities so we can shift to the social practices that will allow us to use natural systems sustainably. It is no exaggeration to say that the survival of human civilisation depends on our ability to respond to this challenge. So it is not just a challenge to the scientific community, but also a challenge to our political institutions. Wouldn't it be wonderful if some of the leaders we pay so generously to think about our long-term future actually did it, instead of squabbling about such trivia as petrol prices, or scheming to manage the increasing diversion of public resources to meet the desires of powerful interest groups? Then we might even get support for sustainability science. But don't hold your breath – we can't expect the future of human civilisation to get much of a go in an election year.

NEITHER SCEPTICAL NOR AN ENVIRONMENTALIST

Published in The Skeptic, *2003*

I was asked to review the controversial book by Danish statistician Bjorn Lomborg, called *The Skeptical Environmentalist*. I concluded that he was neither sceptical nor an environmentalist in any normal sense of those words. A better title might have been *The Gullible Economist*. In fairness, the book has some good points. It correctly points out that some environmentalists are either selective in their use of evidence or not rigorous in their thinking. If Lomborg had applied the rigorous thinking he advocates, he would have extended that criticism to some industrialists, many economists and most politicians – but he didn't. He is very selective in his scepticism. As another example, he analyses the limits of global climate models, but accepts uncritically the much shakier claims of economic models.

In other cases, Lomborg is just wrong. Like most economists, he misrepresents the important report commissioned nearly 40 years ago by The Club of Rome, an international group of business leaders and politicians. Lomborg claims that this report, *The Limits to Growth*, predicted that we would run out of resources. It actually said that limits to growth would be reached within a hundred years if *all* of the trends of increasing population, resource use, industrial production, agricultural output and production of waste were to continue, before showing that it is possible to redirect development onto a sustainable path. Lomborg claims that the IPCC climate projections are 'worst case' scenarios, when the scientific panel said its estimates could be wrong in either direction.

Lomborg lists the broad litany of environmental problems: 'forests are shrinking; water tables are falling; soils are eroding; wetlands are disappearing; fisheries are collapsing; rangelands are deteriorating; rivers are running dry; temperatures are rising; coral reefs are dying; and plant and animal species are disappearing'. He then claims to have refuted them. In fact, *almost all of those statements are true for Australia*. Most of them are also true globally. The second national report on the state of the environment, prepared for the Australian Government by an independent committee, noted some good signs before stating that the environment 'has improved very little since 1996, and *in some critical aspects has worsened*', blaming the compounding pressures of growing population and increasing material demands per person.

The third UN report on the *Global Environmental Outlook* found '*indisputable evidence of continuing and widespread environmental degradation*'. It said policy measures have not been able to counter the pressures of unsustainable consumption levels in rich countries and increasing numbers of desperately poor people in the developing world. It specifically noted problems of water stress, species extinction, depletion of fish stocks, land degradation, forest loss, urban air pollution in developing countries and increasing greenhouse gas emissions. That is the very depressing picture that comes from *scientific analysis*. Lomborg argues that we are not losing biodiversity because the lowest of the credible estimates for the current rate of species extinctions is only 1500 times the average rate of losing species over the Earth's total history. Biologists like Lord Robert May, President of the Royal Society, quote that same figure as being typical of major extinction events.

The fundamental belief driving Lomborg's argument is that 'it is imperative that we focus primarily on the economy' – the environmental equivalent of the discredited trickle-down model of economic development. It suggests our environmental problems will be solved if we get rich enough. The book has a graph showing wealthier countries are more likely to have clean environments. While there is a rough correlation, the actual data reveal that some nations with a gross domestic product (GDP) below US$1000 per head have better environmental quality than others with over US$20,000 per head. So there isn't a simple link.

The second problem is a logical fallacy. A similarity between two changes doesn't mean one is causing the other. The number of lawyers in Australia is growing and so is the number of drug addicts, but there is no reason to suggest the changes are linked. Even if there is a connection, as in the fact that taller people tend to be heavier than shorter people, it doesn't necessarily suggest a policy response; you won't become taller if you put on weight! If the best way to clean up the environment was to increase the rate of economic growth, it would actually be sensible to trash the environment to get rich because we could then afford to clean up the mess. That has been our approach. It is now clear that some environmental problems are effectively irreversible. No amount of wealth will bring back an extinct species, or restore saline land on any human time scale.

Politicians and industrialists like to believe that things are getting better. So Lomborg's claims were hailed by the usual suspects: those on the right of the political spectrum and the ecologically illiterate. But the scientific evidence is clear: we have very serious problems which demand urgent attention. As the UN Environment Program (UNEP) report *Global Environmental Outlook 2000* (*GEO2000*) said, the present course is unsustainable; 'doing nothing is no longer an option'.

The Australian Bureau of Statistics' (ABS) 2002 report *Measuring Australia's Progress* found all the economic indicators for the 1990s were positive, but social indicators were mixed with worrying trends. And *all the environmental indicators measured by the ABS were getting worse*, except for urban air quality. So economic growth is not delivering better social conditions and environmental improvements. In fact, our economic growth is running down our natural and social capital. That is no basis for a sustainable future. As the first national report on the state of the environment said, achieving our stated goal of sustainable development requires the integration of ecological thinking into all our social and economic planning. A naive faith in the magic of the market or the power of growth is no substitute for considered policies that nurture our natural and social systems. Propaganda units like the Institute of Public Affairs fund the travel of people like Lomborg to muddy the water and obscure the harsh reality that we are not using our natural resources sustainably. The *facts* show that we desperately need a new approach. Trusting business and the magic of markets has actually caused the problem; it cannot solve it, even in principle.

UNBELIEVERS: A REVIEW OF *THE GREATEST SHOW ON EARTH* BY RICHARD DAWKINS, BANTAM PRESS, 2009

The Monthly, September 2009

You wouldn't believe me if I claimed that about half the people in a rich modern country still thought the Earth was flat, or that the Roman Empire never existed. It would be a ridiculous assertion. Yet we know that almost half of all US adults believe that humans did not evolve from other species, but were created by a divine being less than 10,000 years ago. The same is true for Turkey, now being considered for admission to the European Union. As Richard Dawkins puts it in his new book, believing that the world is less than 10,000 years old when it is actually about 4.6 billion is equivalent to thinking that North America (or Australia, for that matter) is less than 10 metres across! It is wrong by a factor of about a million. We all have things we don't understand and others that we get wrong, but to be that far out is, as Dawkins writes, to be 'deluded to the point of perversity'. It requires denying or being ignorant of whole fields of science; not just biology, but also physics, chemistry, geology, cosmology and archaeology as well as history.

Dawkins, one of the world's leading science communicators, holds the Chair of Public Understanding of Science at the University of Oxford. Infuriated by the facile nonsense of 'creation science' and its recent re-badging as 'intelligent design', he has assembled in this new book the evidence for the Earth's unimaginably great age and the time that has given for species to evolve through natural selection. It is, as the cover claims, a brilliant tour de force, drawing together an impressive range of science. It reminds us that we are surrounded by a huge range of examples of evolution driven by human selection: our food crops; milking cows; racehorses; the variety of garden flowers; and the plethora of different dog breeds, all evolved from their common ancestor the wolf. In discussing these examples, Dawkins reminds us that there are always trade-offs in selective breeding; we emphasise one characteristic at the expense of others. He notes that the breeding of racehorses for speed makes their legs fragile and susceptible to injury, especially when jumping. In that sense, the recent spate of horses dying in Victorian jumping races is a predictable consequence of selective breeding for speed at the expense of bone strength.

The book also documents a wide variety of examples of natural selection in action, ranging from the reducing weight of elephants' tusks in Uganda to the way species respond to changing dietary or breeding opportunities. More amazing are the changes observed in laboratory experiments that followed evolution through 45,000 generations of bacteria, equivalent to about a million years of human history: longer than our entire existence as an identifiably different species. To drive home the point, Dawkins also draws together the evidence from the fossil record, genetics and molecular biology, all showing that there is no rational basis for denying evolution; it requires serious, industrial-strength delusion.

So why did he feel the need to power up his intellectual sledgehammer to crack a nutty delusion? He says he was provoked by the studied refusal to accept this huge body of scientific knowledge. That denial is usually associated with religious fundamentalism – mostly Christianity in the US, mostly Islam in Turkey. We had a brush with 'creation science' in Queensland 25 years ago, when Lin Powell as Minister for Education told science teachers that they could teach evolution as a scientific theory, but they should teach creation as the origin of species! Most teachers ignored this ridiculous advice, but a small group of fundamentalists at one state high school in Brisbane took it as the green light to push their creationist agenda. We started to find students coming to study science at university, convinced that evolution was a dubious theory peddled by militant atheists to erode religious faith!

It was quite easy to debunk 'creation science' for its cherry-picking of data, its dishonest misrepresentation of reputable scientists and its willingness to ignore whole bodies of inconvenient truths. Interestingly, given his recent embarrassing contribution to the climate change debate, geology professor Ian Plimer played an honourable and feisty role in the debate; his book, *Telling Lies for God*, brutally exposed the dishonesty of 'creation science'.

While there were obvious deficiencies in arguments for creationism, I thought that it had a more fundamental flaw. It was based on an explanation that was not negotiable and would not be affected in any way by evidence or observation: the *a priori* assumption that the book of Genesis was literal truth. That position is the very antithesis of science. I reflected at the time that this nonsense might have been inadvertently promoted by a style of science teaching that presented science as a body of knowledge, a set of eternal truths supported by appeal to the authority of the scientific community. The 'creation science' group claimed God was on their side, giving them even greater authority for their version of the 'truth'. But science is not a body of permanent knowledge, it is a *process* of trying to explain the complex world and advance our understanding. The acid test of a scientific theory is whether it makes predictions which can be tested by observation or experiment, whether it could be falsified by evidence. We can be very confident about the science of climate change because we are now seeing the effects that were predicted 20 years ago. Still, what we call 'knowledge' is always provisional. Research is constantly showing that what was previously accepted is an incomplete understanding, or even wrong. The most dramatic example in my adult lifetime was the discovery of sea floor spreading in the mid-Atlantic. This transformed the naive populist idea of 'continental drift' into the respectable science of plate tectonics, almost overnight.

So what might the impact of this book be? I doubt it will convince the creationists because their position is not based on logical argument. I see a parallel in those who are still denying climate change; as each spurious argument is refuted, they concoct a new objection to avoid the inconvenient reality that we need to change our energy use quite radically. The book will be useful to those seeking reassurance that their scientific understanding is solid. They may need this encouragement. Dawkins notes a worrying trend in the UK for fundamentalist Christians and Muslims to object to the teaching of evolution. It is not yet as bad as the situation in the US, where textbooks are doctored to pander to this sort of dogma. So a generation of young people is being deprived of an opportunity to understand one of the most powerful ideas in science. More importantly, they are also cut off from realising that knowledge can free us from a limited view of the world and empower us to shape a better future.

The insights of science have gradually transformed us from believing we were the summit of creation in the centre of the universe. We now see ourselves as the dominant predator on one planet circling an ordinary star in a remote corner of an undistinguished galaxy. Some are understandably unsettled by the change and long for a return to an earlier, simpler world. The rise of religious fundamentalism can be explained by this wish to retreat to an earlier epoch of certainty. But civilisation will only survive if we accept that we depend on complicated ecological systems which we do not yet fully understand. Retreating to an earlier, simpler world could offer short-term comfort, but would condemn us to the dismal prospect of an impoverished future in a degraded environment. That is why we must not be complacent and assume that denial of evolution only happens in the US and Turkey. We should see defending the process of scientific inquiry and introducing our children to it as investing in our common future.

ENERGY FOR SUSTAINABLE FUTURES

Based on a chapter in Ten Commitments, *Steve Morton, David Lindenmayer, Stephen Dovers and Molly Harriss Olsen (eds), CSIRO Publishing, 2008*

Economic development in the 20th century was fuelled by plentiful cheap energy. It has been clear for decades that the energy outlook for this century is totally different. There is disagreement about the peak of world oil production, with optimists thinking it might still be up to ten years away, while pessimists think it has already passed. Whether optimists or pessimists are right, there is no escaping the conclusion that the age of plentiful cheap petroleum fuels is ending; the scientific basis for 'peak oil' was established more than 50 years ago. So the energy source which now powers almost all our transport will certainly become more expensive. Depending on the politics of the Middle East, oil supplies may also be limited. The near-term future will require a new approach to transport. Public subsidies have encouraged road freight rather than rail and coastal shipping, while inept urban planning has encouraged single-person car use for city trips. These wasteful practices are squandering limited petroleum fuels.

The second challenge for future energy use is global warming. The science has now been refined to the point where there is no serious dispute about the human influence on climate. Australia rejoined the international community in 2007 by ratifying the Kyoto Agreement. The Bali Conference set out a process for developing a future global treaty that can potentially accommodate all major polluters. The issue is very urgent. It is no exaggeration to say that the future of human civilisation is in the balance. Climate change doesn't just have short-term effects on human societies or our economic prospects. The Millennium Assessment Report warned that we are losing species at an accelerating rate as the driving forces of habitat loss, introduced species and chemical pollution are supplemented by climate change. This report forecast we could lose between 10 and 30 per cent of all mammal, bird and amphibian species this century. These are alarming consequences that demand a concerted international response.

So securing our energy future requires an integrated approach. The first step must be setting a science-based emissions reduction target and committing the resources and policies to achieve it. The science shows that the world's greenhouse gas emissions must peak by 2015 and then decline steeply so as to be no more than 40 per cent of the present level by 2050. To achieve this global target while allowing for improvement in material living standards in poorer countries, the IPCC has called for industrialised nations to reduce their emissions by 25–40 per cent by 2020 as a first step toward cuts of 80–95 per cent by 2050. Some government and industry sources have promoted much weaker targets, effectively putting short-term economic goals ahead of Australia's obligation to the rest of the world. We must set a serious science-based target and develop a concerted plan to achieve it. This will require courageous action from all levels of government to develop cleaner energy supply and much more efficient conversion of energy into the services we need.

To reduce the amount of carbon dioxide we put into the air, we must use cleaner fuels and use them more efficiently. Using coal-fired electricity to heat water or cook, rather than burning gas, puts about four times as much carbon dioxide into the air. Renewable energies, such as solar or wind power, release very little carbon dioxide, so they should be the preferred option. Where it is impractical in the short term to scale up renewable electricity to replace coal-fired power, gas should be used as a transition fuel.

Equally important, we must convert energy more efficiently into the services we want. Nobody actually wants *energy*; we want hot showers and cold drinks, the ability to cook our food, wash our clothes and move around. Most of the technology we use is very wasteful. The European Union now has a target of cutting energy use by a quarter by 2020, and some countries like the Netherlands have more ambitious aims. There is no reason at all to be less ambitious than the EU; since Australia hasn't taken many of the easy cost-effective actions already adopted in Europe, we should be able to achieve greater savings in the near future. Saving energy is often much cheaper than buying it. *The Natural Advantage of Nations*, published by The Natural Edge Project, is a weighty tome outlining many specific examples of innovations that make economic sense as well as reducing the environmental costs of meeting human needs. It gives a number of case studies showing that improving efficiency makes business sense. At the household level, if domestic appliances are more efficient, people save money as well as slowing climate change. So we should move immediately to set minimum efficiency standards for office equipment, domestic appliances and industrial processing, reflecting world's best practice. We should also improve building standards for houses and offices, again reflecting best practice in design and orientation to reduce energy demand.

Thirdly, we should set targets for renewable energy in the same way that progressive nations in the northern hemisphere have done. We could aim at generating 25 per cent more electricity from renewables by 2020 than we are currently, and 50 per cent more by 2030. These are realistic targets based on existing technology. The change need not involve significant price increases. More than 15 years ago, a Commonwealth report estimated we could get 30 per cent of our electricity from renewables at no significant extra cost. The clean energy supply technologies have improved dramatically since then, despite meagre funding and limited political support compared with the huge sums lavished on 'clean coal' and nuclear delusions. The 2006 report *Bright Future* showed that we could still get 25 per cent of our power from a mix of renewables by 2020, despite 15 years of inaction since the 1992 report. This strategy would be better for employment and the economy generally than the present approach. The 2003 government report *National Framework for Energy Efficiency* estimated that domestic, industrial and commercial energy use could be cut 30 per cent using measures that would repay the initial investment in less than four years. That approach would create more than 10,000 jobs in activities such as retro-fitting buildings, installing solar hot water systems and replacing inefficient equipment, mostly in regional Australia. Efficiency measures and a real commitment to renewable energy would employ about as many people as the entire workforce of the coal industry. As this essay was being edited for this book, a new US study concluded that the entire world could be powered by renewables by 2030 at acceptable cost, using technology that exists and is proven cost-effective now.

In *A Clean Energy Future for Australia*, Hugh Saddler and his co-authors concluded that solar hot water and improved efficiency could be used to reduce electricity demand in 2040 to 14 per cent below the 2001 value, despite anticipated population growth. This is a crucial point. Studies that assume continuing growth in energy use often conclude that new renewable capacity cannot be built fast enough. Solar hot water is the most obvious cost-effective way of reducing electricity demand, since the time to recoup the capital cost almost anywhere in mainland Australia is less than the guarantee period for modern equipment. It makes sense to mandate solar hot water now for all of mainland Australia except those few sites where solar access is limited by other buildings.

There are always winners and losers from major policy changes, so we should develop transitional strategies to handle the structural consequences. Achieving change on the necessary scale requires price signals, appropriate regulation and a process of social learning involving the whole community. The obvious way to fund the transition is to phase out the huge current subsidies of fossil fuel supply and use. Various studies estimate the annual public subsidy of fossil fuel supply and use in Australia to be between 5 and 8 billion dollars, without allowing for the costs of climate change. We should systematically transfer these public funds to the expansion of renewable energy supply technologies and efficiency gains. A high priority should be the restoration of a fund to support research, development and demonstration projects in these fields.

We should also provide financial incentives to encourage low-carbon approaches to meet our material needs. There is still a robust debate about the relative merits of carbon taxes, with the revenues used to fund development of clean alternatives, or an emissions trading scheme. A well-designed cap-and-trade system would be a solid basis for limiting our future emissions, while a poorly designed scheme would simply allow profiteering by the worst polluters. While emissions trading has the potential to harness market forces to achieve savings in economically optimal ways, design of the trading regime will be crucial.

We should begin planning immediately to reduce our transport fuel use, inflated by the long-standing subsidies of road freight, the encouragement of single-person car use for urban trips and vehicle efficiencies that are very poor by international standards. As well as moving to set serious standards for vehicle efficiency and phasing out the subsidy of road freight, we should begin to invest in world-class public transport systems for all major urban areas, rather than squandering huge sums on dinosaur road schemes. The twin forces of climate change and 'peak oil' both demand a move to improve dramatically the fuel-efficiency of urban transport. Future planning should also emphasise compact urban villages with everyday needs within walking or cycling distance to reduce the transport task. Such cities would provide a better social environment and improve community health by promoting physical activity in more natural surroundings, as well as meeting energy goals.

As discussed above, some changes require regulation: efficiency standards and targets for clean energy supply are obvious examples. Governments will have to be modelling good practice by ensuring that they set the highest environmental standards for public buildings as well as for energy and resource use by government and public authorities. The sustainability reform agenda clearly needs an institutional base to work through the Council of Australian Governments (COAG), taking a whole-of-government approach at all levels. The Rudd Government has a policy commitment to establish a National Sustainability Commission which would develop a sustainability charter, setting out broad principles of a desirable future as well as specifying science-based targets for key areas such as energy, greenhouse gas emissions, water use, transport, biodiversity, buildings and urban planning. There was no sign of this commitment being implemented as this essay was finalised, more than two years after the 2007 election. Such a commission would be a vital step toward implementing the reforms needed.

Finally, a low-carbon future will only be politically sustainable if it is developed through a process of public involvement, along similar lines to the approach used recently in Sweden to arrive at their future energy strategy. A community conversation to develop our future energy strategy should be a high priority.

To summarise, the two big challenges of 'peak oil' and climate change demand a different approach to energy supply and use. A concerted response strategy is environmentally essential, technically possible, socially desirable, economically achievable and politically preferable to the alternative of waiting until there is massive social and ecological disruption. It should not need saying, but a future sustainable society clearly must have stabilised both its population and its overall resource use, including its energy consumption, at levels consistent with the limits of natural systems. This is a radical departure from present thinking and would be widely seen as bordering on heresy, but it is an inescapable conclusion if we are serious about a sustainable future.

SOCIAL AND ENVIRONMENTAL IMPACT ASSESSMENT: A CASE STUDY OF THE HYDROGEN ECONOMY

Conference presentation, 2004

Until 30 years ago, there was no tradition of assessing the social or environmental impact of changes. In 1974 the Whitlam Government passed the *Environmental Assessment (Impact of Proposals) Act*, making it possible for the first time for an Australian government to require an inquiry into the possible consequences of a proposed development. Various countries introduced more comprehensive assessment systems; for example, Sweden set up a Secretariat for Futures Studies in the late 1970s and even the US Congress established an Office of Technology Assessment to advise decision makers on the social and environmental consequences of new proposals. In Australia, the Hawke Government set up a Commission for the Future in 1984. Its goal was to assess social and other impacts of developments in science and technology, in the hope of helping the community make more informed decisions. I was its director in 1988. It did significant work across a range of issues, from advances in genetic technology and the developing understanding of global climate change to the impacts of an ageing population. The enthusiasm for a mindless acceptance of market forces in the 1990s saw the Commission's funding wound down and finally discontinued. We now require major developments to produce environmental impact assessments, but these are produced by the proponent and so inevitably find that the environmental impacts are acceptable. We have no requirement for social impact assessment, even though the social impacts of new technologies are often more serious than the environmental effects. The present dominant ideology is a naive faith that the market will handle these issues.

The case for the future hydrogen economy rests on three foundations: depletion of petroleum reserves, global change and the need for a secure, equitable world. So I am an unashamed advocate of hydrogen futures. There are, however, some serious issues that we have to confront if those futures are to eventuate.

About 70 years ago, King Hubbert used statistical data on US oil discoveries and associated production to predict that US oil output would peak about 1970. It did, leading to a change in the relationship between petroleum producing nations and those which use oil. The 1970s oil 'shocks' dispelled the myth of infinite resources, causing significant policy changes in many Northern Hemisphere countries. Hubbert's technique was being used in the 1970s to estimate that world oil production would peak about 2010, plus or minus ten years. That is still the best estimate: there are optimists who think the peak might be as far away as 2020, while there are pessimists who think it happened in the year 2000! Whoever is right, there can be no escaping the fundamental geological truth that we are using petroleum much faster than it was produced naturally, so it will not be plentiful for much longer. Some analysts think the struggle for the remaining oil is already under way. Professor Gretchen Daily of Stanford University posed a rhetorical question to a forum in Sydney last year: 'How concerned would the US administration be about human rights in Iraq if it only had 10 per cent of the world's broccoli?'

Military and other geo-political issues aside, most decision makers are still in denial about the approach of the peak in world oil production and the consequent need to change our transport systems, which are still based on the presumption that fuel will continue to be plentiful and cheap. While fuel prices in Australia are much higher than in North America, they are much lower than in Europe. In fact, Australians pay more per litre for beer, cask wine, milk, orange juice or even bottled water than they do for motor spirit! Cheap fuel leads to profligate use, with many urban commuters driving long distances as the sole occupant of large and inefficient cars. The situation is being worsened by the current fad of buying large four-wheel-drive vehicles to cope with the uneven terrain of suburban streets. A recent survey in Australia found that the dominant reason for buying these urban assault vehicles is that drivers no longer feel safe on the road in sedans, given how many large four-wheel-drive vehicles are on the road! So there is an urban arms race, leading logically to armoured cars, tanks and Humvees on the street.

There are alternatives to oil as a transport fuel, but most of them pose their own problems. In the short term, it is relatively easy to envisage natural gas being used. Liquefied petroleum gas (LPG) is already the fuel for most taxis in Australian cities, while Brisbane City Council decided a few years ago to move its bus fleet from diesel to compressed natural gas. The problem is that gas is also a finite resource. While the known reserves of gas are much greater than the reserves of oil, using gas for all transport purposes would roughly quadruple demand, so the change would only buy a decade or two of breathing space. Many see gas as a bridge to a sustainable future, taking advantage of that breathing space. That possibility leads to a questioning of the current enthusiasm to export gas. If all the large projects currently under discussion were to materialise, Australian gas exports would increase from the present 7 million tonnes a year to about 40 by 2020. The sale of the gas would produce short-term economic benefits, but it may be doing future generations a disservice by selling a resource they might need.

A second group of fuel alternatives contains those that can be produced sustainably from plant material. Australia has produced ethanol from sugar since the 1930s, while Brazil and the US also produce large quantities from sugar and maize respectively. There are three problems with ethanol. The collection and processing of crops like sugar requires significant amounts of transport fuel, so the energy benefits are small or – in the view of some analysts – possibly negative! The second problem is that growing sugar leads to other environmental problems; in Australia, pollution of the waters around the Great Barrier Reef is a consequence. It is now accepted in the scientific community that the outbreaks of crown-of-thorns starfish are a relatively recent and direct consequence of the effects of run-off into the reef lagoon of nutrients from agricultural activity. The third problem is the ethical dilemma of whether it is appropriate to use food-growing land to produce transport fuel in a world where millions go hungry. The scale of the potential contribution is also limited; converting Australia's total sugar production to ethanol would meet about 10 per cent of transport fuel needs.

Methanol looks promising because a 1979 CSIRO study found that we could produce enough of it from rapidly growing trees (biochar) to supply all of Australia's transport fuel, but only if it were possible to use an area about the same as that now devoted to all agricultural purposes! So plant-based alcohols may be a useful supplement of petroleum fuels, but they are unlikely to be produced on a scale sufficient to be a replacement.

From time to time, other hydrocarbon resources such as oil shale and tar sands are promoted as the solution to the problem of transport fuels. There is no doubt that it is technically possible to produce synthetic fuels from these sources. A site at Glen Davis, not far west of Sydney, produced oil from shale when I was young. The problem with oil from shale and tar sands is the economics, but the economic issue is actually a manifestation of a deeper and more fundamental obstacle. Let me remind you of the history of oil shale, as a cautionary tale. In the early 1970s, when the price of oil was less than US$2 a barrel, the entrepreneurs with shale oil concessions said that the product would be economically viable if the price were ever to reach $5. When it did, the break-even point had drifted up to seven or eight dollars a barrel. I will spare you all the intervening steps and remind you that a current price of around US$50 a barrel has not been enough to make oil production from shale viable. There is a fundamental explanation: oil shale is a low-grade resource, typically containing about 100 kilograms of oil equivalent per tonne of rock. It takes large amounts of energy to dig up the rock, crush it and process it to obtain the hydrocarbons. Energy analysis by Gerald Leach 25 years ago suggested that the energy used to process typical shale deposits is about the same as the energy content of the product. So, in energy terms, the process is effectively paying one lot of men to dig a hole and another lot to fill it up. However high the price of oil goes, the economics will always be dubious. Glen Davis was an unusually rich deposit, yielding about 350 kilograms of product per tonne and proving profitable. Many shale deposits are very large, with huge amounts of potential hydrocarbons, but the low grade of the resource makes the economics questionable. Processing shale on the scale needed to make an impact on fuel needs would also cause serious environmental problems; Greenpeace regularly campaigned against the proposed oil shale pilot plant at Gladstone for that reason.

So one solid argument for hydrogen as a fuel is essentially that it appears the only realistic solution to the problem of a world dependent on plentiful transport fuels, as the peak of world oil production leads to scarcity and higher prices for petroleum products.

The second reason for supporting hydrogen is the impact of global climate change. While carbon dioxide levels in the atmosphere have varied between about 180 and 280 parts per million for the last 500,000 years, the figure is now about 380. This is a direct result of burning huge amounts of coal, oil and gas since the Industrial Revolution. The Earth is now about 0.7 degrees warmer than it was a hundred years ago, with consequent changes to rainfall patterns, plant growth, distribution of animal species, sea levels and the frequency of severe events like storms, floods and droughts. The city of Perth now faces severe water problems as a direct result of climate change. Where the average run-off to the city's water supply reservoirs for the first 75 years of last century was about 370 megalitres, the average figure since 1998 is 115. The state government has serious plans for a desalination plant and tapping a deeper aquifer to meet the city's needs. All the projections suggest the situation will get much worse. The IPCC, the UN's advisory body, gave in its Third Assessment Report a range of possible future outcomes, depending on the pattern of future fuel use and taking account of uncertainties in the science. The worrying 'bottom line' is that the most optimistic future, based on a rapid phasing out of fossil fuels and the best interpretation of the scientific uncertainty, still involves a further 1.5 degrees increase in average global temperature, with associated changes in other outcomes influenced by temperature. That is why the governments of most industrial nations have agreed to begin a process to reducing emissions of carbon dioxide. The first step is the Kyoto Protocol, still unfortunately being obstructed by short-sighted politicians in the US and Australia, but now likely to be ratified and enter into legal force early next year. Some nations are looking well beyond the first timid steps in the Kyoto agreement; for example, the UK recently adopted a target of reducing its carbon dioxide emissions by 60 per cent by the year 2050. Even if petroleum reserves were unlimited, climate change would be requiring us to look seriously at ways of reducing its use for unnecessary purposes. Future generations will find it difficult to believe that we drove alone in commuter vehicles, and will probably be startled to learn that we actually fuelled vehicles and raced them around a track just to see which was the fastest.

The International Geosphere-Biosphere Project recently released a major report on its decade-long scientific study. *Global Change and the Earth System: A Planet Under Pressure* was published by Springer-Verlag this year. It points out that global change is more than just changes in the climate. It is real, is happening now and in many ways is accelerating as the multiple interacting effects of human activity cascade through the natural systems of the Earth. In addition to carbon dioxide levels, several other parameters of the Earth system are now well outside the natural variation observed over the last half million years or more. We also know that the dynamics of the natural systems of the Earth are characterised by critical thresholds and abrupt changes when those thresholds are exceeded. So, the report warns, it is entirely feasible that human activity 'could inadvertently trigger changes with catastrophic consequences'. The analysis leads to an obvious conclusion: 'Dramatic increases in energy efficiency, decarbonisation and the development and utilisation of new sustainable energy technologies, such as a hydrogen-based energy system, are needed.'

So the second leg of the case for hydrogen is the need to move away from the energy sources that are changing the global climate and threatening our future.

The third leg is the need to move toward a more equitable world if we want a sustainable future. I don't believe that entrepreneurs in the OECD countries can be secure doing property deals on mobile telephones as they speed around in large cars in a world where the majority of humans have never ridden in a car, never made a phone call and never owned property. The division between the 'haves' and the 'have nots' is widening all the time. In 1980, the richest 20 per cent of the world had 70 per cent of the wealth and the poorest 20 per cent had 2.3 per cent: a ratio of about 30:1. By 1995, the ratio was 60:1 and today it is 75:1. That trend cannot lead to a secure future. The dominance of US media has also made many people in the poorest countries acutely aware of the relative material comfort of the wealthy nations. An increasingly inequitable world is very likely to be an increasingly insecure world, with literally millions of people from poor countries risking their lives to make their way to North America, Western Europe or the relatively affluent countries of the Pacific Rim: Japan, Australia and New Zealand. It is possible to imagine social and political solutions to basic needs; for example, the UN Development

Program (UNDP) has estimated that the entire developing world could be given adequate nutrition, clean drinking water, reasonable shelter, basic education and health care for about 5 per cent of the global military budget. But there is no prospect even in principle of extending to the developing world the sort of access to transport energy we take for granted. It has been estimated that if the entire world used oil at the rate of Australians, and it could be pumped out fast enough, the entire global reserves would be exhausted in about a year! So the only prospect of more equitable access to energy, arguably a prerequisite for a secure and peaceful future, involves the development of energy technologies based on plentiful resources. That makes a compelling case for promotion of hydrogen futures.

While I am an enthusiastic supporter of hydrogen futures for the reasons I have summarised, I am also aware of some problems. The wise US journalist HL Mencken was quoted as saying that every complex question has a simple answer, but it is always wrong! It is never possible to change only one thing in a complex system, because other changes always follow. So we should temper our enthusiasm for hydrogen futures with an awareness of the outstanding issues.

There are some technical problems still to be solved before we can be confident of having reliable energy systems based on local conversion of hydrogen using fuel cells. The papers presented at the recent international conference on hydrogen and fuel cell futures in Perth make me very optimistic that these problems are soluble. In general, we can be reasonably confident on the historical evidence that problems that are *purely technical* can be solved if sufficient resources are devoted to the task.

The technical issues overlap with the economic questions. Both technically and economically, the most attractive way to produce large amounts of hydrogen is to use fossil fuels like coal or natural gas as the feedstock. But serious analysis suggests that the natural gas – hydrogen – fuel cell approach is approximately greenhouse neutral, while starting from coal actually worsens the greenhouse impact of transport. If we want the hydrogen economy to reduce our assault on the climate system, we will need to produce the hydrogen by using renewable electricity to split water. Turning solar or wind energy into hydrogen is a way of storing those intermittent sources, so it has obvious appeal. The problem is that the present technologies make hydrogen from renewable electricity economically unattractive. So the prospect of hydrogen fuel cell futures eventually being based on clean renewable energy hinges on development of less expensive ways of harnessing those energy forms. There are promising studies under way of geothermal energy from hot dry rocks, of wave energy and of new generations of the established technologies like solar cells and wind turbines. At this time, however, it seems likely that these 'clean' alternatives will mean much more expensive transport energy.

The third issue is the environmental impact of large-scale hydrogen production. As I have already discussed, the hydrogen route may not solve the problem of climate change. Producing solar cells, or wind turbines, or any other renewable energy devices on the scale needed to supply enough hydrogen for the global transport task will have environmental impacts. A more serious issue is the uncertain impact on atmospheric chemistry of large-scale use and inevitable release of hydrogen. As the lightest molecule, hydrogen is notoriously difficult to contain. A study by a group of researchers at CalTech, published in the journal *Science* in 2003, calculated that a future global hydrogen economy could result in 60–120 million tonnes of hydrogen being lost each year into the global atmosphere. We just don't know what the impact would be. An optimistic view would hope that it would just react with oxygen to form water and return to the oceans, or that it would be absorbed in ways that don't cause unforeseen problems, but we cannot be certain. So we should be doing the research now. As mentioned earlier, Nobel laureate Paul Crutzen has pointed out that we were just lucky that CFCs were based on chlorine rather than bromine, and so thinned the ozone layer rather than destroying it entirely! Some researchers have also pointed out that the lightness of hydrogen makes it likely that significant amounts would drift into the stratosphere and possibly even be lost from the Earth system. They argue that taking water from the oceans to transfer hydrogen into the atmosphere might, in time, lead to reductions in sea level. I could make the frivolous observation that this might redress the rising sea level caused by global warming, but we would obviously have to be very confident about our sums if we were to address this as a conscious strategy! The serious point is, once again, that this is an issue that should be studied. Before we embrace the hydrogen economy, we need to have done enough research to be confident we are not getting out of the climate change frying pan into some unknown environmental fire.

The fourth issue is the social and political impact of working toward a hydrogen future. The time, intellectual effort and other resources expended on one area of research and development inevitably reduces the capacity to work in other fields. Some well-intentioned people see the nub of the transport problem not as oil and the internal combustion engine, but the inefficiency of the average vehicle. They have a point that cannot easily be dismissed. Imagine giving a group of engineering students the task of designing a transport vehicle to carry a fragile payload, typically between 50 and 100 kilograms. If they produced a design that weighed more than a tonne, you would almost certainly recommend they review their career options, possibly steering them toward a future for which numeracy would not be important, like politics. It is easy to imagine a vehicle like the Amory Lovins hypercar, weighing about 250 kilograms instead of five times as much and as a direct result using only one-fifth the fuel. Making such vehicles the norm would make the limited oil reserves last five times as long and dramatically reduce emissions. Others go back one step further and see mobility as a response to poor urban design. Again, these people have a point. The reason we move into cities is to access the wider range of services that are available there. We don't move to cities so we can spend hours travelling to access those services. So a fundamental priority should be improved urban design, making the services people want easily accessible, rather than accepting the inadequacy of current design and flattening increasing fractions of our cities to allow people to drive by themselves for hours in pursuit of the services that are now inaccessible.

I concede the relevance of these criticisms. We should, of course, encourage improved urban design and smaller, more efficient vehicles. But these measures have long time lines. Much of the structure of the cities of 2030 is already in place. Most of the vehicle fleet of 2015 is already on the roads. So we need to be pursuing cleaner energy systems as well as trying to develop better vehicles and putting greater effort into urban planning. Like the other problems I have discussed, these issues should not deter us from working on hydrogen fuel cell futures; they simply provide the wider context within which we should be doing that work.

I have no doubt that we should be working toward sustainable futures. That is a moral imperative. It is indefensible to be developing futures that we know cannot be sustained, producing inevitable problems for future generations. As the second report in the UNEP series on the Global Environmental Outlook said, our present approach is not sustainable, so doing nothing is no longer an option. A sustainable future will be one in which we are not depleting the resources future generations will need, are not doing serious damage to natural systems and are moving toward an equitable and secure world. So, market-led wealth generation and government-guided technological change has to be supplemented by a values shift towards a new global vision marked by equity and durability.

I believe we should be looking at strategic goals, like stabilising the population and eliminating hunger. That doesn't require technical advances, it simply requires a more equitable distribution of the 2 kilograms of food per person per day we now produce, rather than a market approach in which those who cannot afford food go hungry, while land which formerly grew cheap food for subsistence living in Africa is now increasingly being used to grow flowers to be air-freighted to rich consumers in the developed world. We should also be aiming at a dematerialisation of society. Some European nations have now adopted the goals suggested by the Wuppertal Institute of reducing energy use to a quarter of the present level and reducing material use to 10 per cent of the present level; they see those as realistic targets.

But above all else, we need a values shift, perhaps away from *Homo sapiens*, which is gendered and a link back to our past, towards *Globo Sapiens*, a concept developed by my partner Patricia Kelly for use in education from a term proposed by Pentti Malaska The idea of being wise citizens of the planet recognises that we share it with all other species and that we hold it in trust for all future generations. That means that we need to see the economy as a *means* to service human needs rather than an end in itself, and that we should be committed to *genuine* globalisation rather than the current fad of simply reducing the constraints on corporations.

We have to see that growth in itself is not a solution. The Brundtland Commission pointed out 15 years ago that the two main causes of environmental degradation are extreme poverty in the poor countries and unsustainable levels of consumption in the rich countries. Growth can in principle do something about the first problem; but growth is both in principle and in practice making the second problem worse, and will continue to do so unless we embrace a different sort of growth which is oriented towards human need rather than human greed.

The fundamental problem is still that most decision makers are operating under what could be called the pig-headed model of the world, in which the world is seen like the head of a pig, with the economy a large shape like the face, while society and environment are minor protuberances like the ears. For those who still have that primitive worldview, it actually makes sense to say that the economy is supreme and the minor problems of society and environment can be handled as long as the economy is thriving. If you think about it, the only rational model is one that accepts that *the economy is a part of society*; it is a very important part, but only a part, because we expect from our society a range of services which are not part of the economy, like a sense of place, cultural identity, security, companionship and love. Our societies are totally enclosed within natural ecological systems on which we depend for breathable air, drinkable water, adequate nutrition, a sense of cultural identity, spiritual sustenance and so on.

We tend to behave as though we are not part of natural systems. I remind myself that every molecule of my body was once part of the natural systems of this planet, and indeed every molecule of my body will, in time, once again be part of the natural systems of this planet.

We need to accept that our social and economic planning should be within an ecological framework, that we do need planning and conscious decision making, rather than trusting the magic of the market which cannot, even in principle, represent the interests of other species and future generations. So we need new social institutions, we need new technologies to meet our needs, but above all else we need values for a sustainable future based on the principle of *Globo sapiens* and continuous adaptive management based on social learning.

Social and environmental impact assessment of our current problems and possible solutions is thus not just an operational technique for making better short-term decisions, but a framework which leads to a comprehensive approach.

HUMBLE BICYCLE A WINNER ALL THE WAY

The Weekend Australian, 15–16 June 1990

It's time to peddle a new transport style for our cities. The bicycle is making a comeback. These days we are all being urged to cycle to work and recycle when we get home. Even the Queensland Government, which could hardly be accused of pandering to green activists, has produced a booklet to encourage recycling of such materials as glass, paper and metals. Tastefully printed on recycled paper, the new publication was much in evidence on World Environment Day.

While that day was cloudy and wet in Brisbane, the Sunshine State has since been blessed with the sort of weather which made it possible to endure even the worst excesses of the Bjelke-Petersen regime. With the sun blazing out of a clear blue sky last Monday, thousands of Brisbane people celebrated the holiday by taking to the streets on their bicycles. Arranged by the National Heart Foundation to promote healthy exercise, the Brisbane bike hike attracted a wide diversity of machinery and human forms of propulsion. At one extreme were the earnest types with their aerodynamic helmets and expensive imported cycles. All seemed to be wearing strange black pants which were either very long shorts or very short longs, so improbably tight that I imagined them to have been put on with a spray can for later removal with a tin-opener. They surged away at high speed, forming a rapidly shrinking mobile of bouncing black buttocks.

Behind was a sea of more modest machines and less energetic propulsion units. Some bikes were so shiny and new they might have been bought for the occasion, while others looked as though they had just been found at the back of garages under old ping-pong tables and 17 years of accumulated dust. Light relief came from the occasional unicycle, a few tandems, toddlers on tricycles and an amazing ancient machine that looked as if it had been carved from an old gum tree.

The city rail system allowed cyclists free travel for the day. As a result, the morning train I caught to the city was a gleaming mass of handlebars, spokes and reflectors. Innocent passers-by fought their way onto the train through the sort of barrier which might have been designed for guerilla warfare. The combination of bicycle and train allows people of average fitness to travel quite long distances with speed, comfort and flexibility. You simply cycle to the nearest station, get a train to a stop near your destination and complete the journey on bicycle. Hundreds of people did just that – a vivid reminder of the versatility of the bicycle, by far the most energy-efficient form of urban transport. The bicycle not only uses much less fuel energy than cars, buses or trains, it also uses much less human energy than walking.

I had written off bicycles as suitable only for the young and absurdly fit, but there is a wonderful technological advance that brings cycling within the reach of all. To a lapsed cyclist, the mountain bike is a revelation: a strange amalgam of styles, with chunky tyres, light alloy frame and an awesome range of gears. The combination allows the happy cyclist to go almost anywhere without dismounting: along roads; over footpaths; on gravel paths; and up the steepest hills imaginable. The old days of red-faced straining to avoid the embarrassment of getting off and walking are gone forever. Even someone as unfit as I am can just sit down, relax and pedal slowly upward.

The energy used for cycling per passenger-kilometre is one-third of the figure for walking, one-twentieth of the figure for a bus or train and an amazing one-fiftieth the figure for an average car with one occupant. For urban commuting, this is the operative figure, as the vast majority of cars in peak-hour traffic contain only the harassed driver.

Getting around by bike is surprisingly quick. The hike was 23 kilometres of hilly Brisbane terrain in total. With stops for drinks of water, chewing segments of orange, listening to two different bands and savouring some of the views, I found the full distance took only one and a quarter hours. Many urban commuters spend longer than this sitting behind the steering wheel, getting no exercise and probably doing their blood pressure no good at all. We would be a healthier society if more of us left the car at home and pedalled to work every day, or even just occasionally.

Civic authorities need to provide some encouragement. The new Brisbane City Council has given notice that it will require bike facilities in new city office buildings, and a prominent alderman took part in the bike hike. The real need in most of our cities, however, is for a decent system of cycle paths to separate cyclists from motorised vehicles. This would improve safety, encourage timid cyclists and reassure motorists who understandably worry about erratic bicycles which appear immune to the rules of the road.

A recent report by the Worldwatch Institute referred to cyclists as 'the silent majority'. It is a nice image. By contrast with the noise and pollution of the car, the cycle is clean and quiet. The global production of cycles is about 100 million a year. This is about three times the volume of production of cars; but the price is much lower in terms of resources and energy. Australia actually boasts about as many bicycles as cars, so we are well equipped to take advantage of this healthy form of transport.

In our modern cities people actually pay good money (and some drive their cars significant distances) to pedal stationary bikes for exercise. They would benefit if the exercise actually took them where they wanted to go, rather than just working up a lather in a suburban gym. If we are going to make the city air of acceptable quality and reduce emissions of carbon dioxide, we need to promote cycling for commuters. As a fringe benefit, we will certainly be a healthier society than we are now.

OUR RESPONSIBILITY FOR FUTURE GENERATIONS

Conference presentation 2008

We are deciding the sort of world future generations will inherit. The present outlook is quite bleak. Obsessed with short-term economics, we are systematically neglecting our obligations to consider such issues as resource depletion, environmental damage and social cohesion. The way we are now living in Australia does not appear sustainable and certainly could not be extended to the entire human population. Our moral responsibility to future generations demands a radically different approach.

I have been most engaged in the study of our environmental problems, but we should not concentrate on them to the exclusion of all other issues. In the short to medium term, the impossibility of increasing world oil production to cope with the demand being stimulated will inevitably lead to higher prices, physical shortages and possible serious conflict between different communities. An enthusiastic embrace of a globalised market approach to economics has been remarkably successful in providing short-term material benefits. But it has certainly been accompanied by a widening gulf between the rich and the poor, even if it can't be proven to everyone's satisfaction that the economic approach is actually causing the problem. A world of increasing inequality will not be socially and politically sustainable. The emphasis on short-term economic goals has also directly caused serious environmental problems. At the world level, these have been documented by a series of reports: the UNEP series on the global environmental outlook; the reports of the IPCC; the report summarising the work of the International Geosphere-Biosphere Program; and the UN's Millennium Assessment Report, released in 2005. I have been involved in preparing or reviewing all those reports, which tell a consistent story: we are doing unacceptable damage to the natural systems of the planet.

At the national level, we have now seen three State of the Environment reports. I helped prepare the 1996 report, which concluded that we had very serious environmental problems that will prevent us achieving our stated goal of sustainable development. The second report in 2001 found that all the serious problems were still getting worse. The third report, released in 2006, concluded that there are 'several environmental issues of concern'. But all these reports have failed to galvanise politicians into action and we still lack a concerted response, despite the accumulated weight of scientific opinion. There is a real risk that successive reports on the state of the environment will simply document in detail its continuing decline without achieving any translation of the scientific expertise into public policy responses. The 2006 report gloomily but accurately concluded that 'biodiversity decline will continue because of the consequences of past actions and the time it will take to see the effects of current initiatives', while 'the condition of land, inland waters and coastal lakes' will either remain the same or 'continue to decline for some time'. Yet the then minister hailed the report as a tribute to the government's environmental management!

The IPCC's Fourth Assessment Report repeated the message of

its earlier reports, that human activity is changing the global climate and we should be seriously concerned about this. In two fundamental ways, the report understated the seriousness of our predicament. As a consensus publication, the IPCC report was the cautious, conservative, lowest common denominator of scientific opinion. The need to obtain agreement inevitably results in the editing out of any conclusions that do not command universal support, so even a widespread view that we risk serious non-linear changes to the climate system does not make it into the report. As an editorial in *New Scientist* said: 'This desire for a consensus of certainty comes at a price. The rising tide of concern among researchers about positive feedbacks in the climate system is not reflected in the summary.' The related second deficiency is that the reports must necessarily set a cut-off point for the scientific literature that will be used, so papers published after that date are ignored. In this case, there was no reference to the increasing concern that global warming could slow down the ocean circulation systems. As far back as the 2001 Global Change Science conference in Amsterdam, we were warned that the driving forces of the Gulf Stream were being weakened by climate change. In 2005, as *New Scientist* reminded us after the release of the IPCC report, scientists at the UK National Oceanography Centre 'reported that the Gulf Stream slowed by about 30 per cent between 1957 and 2004'. Far from this being the headline item it should have been, given the horrendous possible consequences if the Gulf Stream were to slow further or possibly even stop, the IPCC report found 'insufficient evidence to determine whether trends exist'.

An even larger worry, given its potential impacts on human societies, is the increasing evidence that warming is destabilising the Greenland ice sheet. Since scientists began monitoring seismic events in this huge body of ice less than 15 years ago, there has been an eight-fold increase in the frequency of serious shocks. So something very serious is clearly happening to this huge ice sheet which is currently perched on the rocks of Greenland. Some scientists worry that even 2 degrees of global warming could trigger the collapse of the ice or its sliding into the ocean; a larger group think that 3 degrees temperature increase would almost certainly have this sort of impact. If the Greenland ice were to slide into the ocean, world sea level would rise about 5 metres. That would have a devastating impact on coastal settlements all around the world. Scientists are becoming much more concerned about the risk we are taking by failing to develop a concerted response to climate change.

As the Millennium Assessment Report concluded, climate change is not just a serious problem in its own right, but also contributes to the other serious environmental problems like the state of rural land, the flow of inland rivers and the loss of biodiversity. I believe the failure to develop political momentum for change is at least partly due to the intrinsic problem of applying science to complex questions. Since we all see the world through the lenses of our experience, our culture and our values, there are different legitimate interpretations of such complex and uncertain issues as whether we risk non-linear changes to the climate system. Fifteen or 20 years ago, there was even legitimate doubt about whether human activity was changing the climate, although that issue has now been resolved to the satisfaction of all but a small number of deniers. This is a reminder that as well as legitimate different interpretations, there will also be scope for what could be called illegitimate views. The report that a large oil company was offering up to $10,000 to scientists prepared to denounce the IPCC report suggested that this sort of corporate bastardry to corrupt science is still happening. However difficult it is, those of us who understand the science have a responsibility to be communicating the seriousness of the situation to our leaders and the community. As long as the media can claim that 'science is divided', however lopsided the division, there will be an excuse for what Sir Humphrey Appleby called 'masterly inaction'. We can no longer afford that strategic approach to our serious environmental problems in general.

It is particularly urgent to develop a concerted response to climate change based on accepting a long-term goal of dramatically reducing our greenhouse pollution, effectively decarbonising our society by 2050, as well as setting a short-term goal like 40 per cent reduction by

2020. With that broad framework to set the cap on our total emissions, it will be possible to develop a trading system that will provide economic incentives for the sorts of changes that are needed. We also need a strong target for increased use of renewable energy. Fifteen years ago, the Commonwealth Department of Resources and Energy (DRE) estimated that we could get 30 per cent of our electricity from renewables by 2020 at no more than 10 per cent extra cost. Despite 15 years of inaction, it should still be possible to get 25 per cent of our power from renewables by 2020. Setting that sort of target would revive the industries that have been closing down or moving overseas in the absence of a strategic approach from the Australian government.

The alternative is bleak. I was challenged in 2007 to think about how bad things will get in the next 20 years if we continue to ignore the warnings from science. It was a very depressing analysis. By 2027 we will have lost a significant fraction of our unique biodiversity from the Alpine areas, temperate bushland and the Great Barrier Reef, where serious bleaching will happen regularly and it is possible the whole reef system will be in terminal decline. Water supply for our cities will have been made more acute by climate change reducing rainfall while urban populations kept growing. The Murray will be in desperate shape and Adelaide's drinking water will be substandard. More of the West Australian wheat belt will have been lost to salinity, more rural land will have been degraded and much of regional Australia will be in economic trouble. The present epoch will be seen as a tragic period of inaction on the crucial issues of climate change, urban planning and infrastructure: public transport, water and clean energy. Expanding the uranium industry will have caused widespread local pollution and proliferation of nuclear weapons around our region, making nuclear war appear inevitable in the context of increasing anxiety about such limited resources as oil.

This analysis was a reminder that we are now effectively deciding the broad direction of our future. Will it be the clean, green road of a sustainable future? Or will it leave our children a dreadful legacy of climate change, radioactive waste and derelict land? This is a critical juncture and we urgently need leadership. As Tony Blair has said, I wouldn't like to be the leader when in 15, 20, or 30 years' time, people look back and say, 'What on earth were they doing at that time?' Do we want our children to blame us when they inherit degraded landscapes, or can only read about the species they will have lost? Our urgent task is to develop lifestyle choices that would be sustainable.

It would be much easier to ignore these difficult issues, to enjoy our material comforts and our wonderful lifestyle – but a sustainable future is clearly a better future. Working for it is our moral duty to the millions of other species we share this country with, and the future generations for whom we hold it in trust. It is entirely possible to develop a new approach, integrating environmental awareness into economic and social decisions. If we work for a global response to climate change, it is still conceivable that we can help to avoid the worst consequences, mainly by encouraging China and India to follow us down the path of clean energy supply and efficient use. If we renounce uranium, our region will be cleaner and safer. If we set migration targets that would stabilise our population, we could achieve that by 2030. If we also stabilise consumption per person, we could reduce our impacts to a point that would enable us to argue realistically that we are living in balance with natural systems. That should be our goal.

SUSTAINABLE FUTURES

Conference presentation 2007

The head of the Tellus Institute in Boston, Dr Paul Raskin, produced in 2004 a postscript to the groundbreaking report *The Great Transition*, developed by the Global Scenarios Group. Raskin's essay, *The Great Transition Today: A Report from the Future*, looked back from 2084 to summarise the changes that made a sustainable society possible. The crucial factor was 'a new suite of values': 'Consumerism, individualism and domination of nature – the dominant values of yesteryear – have given way to a new triad: quality of life, human solidarity and ecological sensibility.' He noted that these would be applied with different weights and shades of meaning in different societies, but they are the underpinning values. In this plausible future, world population stabilises at about 8 billion as a result of improved living standards in poor countries and greater empowerment of women. The economy is seen as a means to social, cultural and environmental ends, rather than as an end in itself. Although the overall scale of the economy in 2084 is seen as being much greater than it is now, the flow of material resources is far less, water is used sustainably and fossil fuel use has been cut dramatically. The societies in this imagined future are much more egalitarian than today's. Participatory democracy exists at the local, regional and global level, enabling the community to make civilised decisions that are politically sustainable because all interested parties have been involved. The essay is an inspiring vision of the sort of world we should be aspiring to build. It demonstrates clearly that the transition to a sustainable future is a moral and ethical challenge as well as requiring profound technological change.

I have argued that we need to become a HEALTHIER society: one that will be Humane, take an Ecocentric Approach, use a Long Time Horizon for planning purposes, be Informed about consequences of our choices on natural systems, be Efficient in its use of natural resources, and be Resourced.

It will be Humane in aiming to develop technologies and approaches that can, at least in principle, be extended to the whole human family, rather than restricted to a privileged minority in a few countries. So it will recognise that we share a common fate with all other humans, a lesson that should have been made obvious by such global environmental issues as ozone depletion and climate change. A humane society will have appropriate safety nets for those who are less privileged and will treat refugees fleeing from troubled countries or poor areas with compassion.

It will have an Ecocentric Approach because it will recognise that we have no future at all unless we maintain the capacity of natural systems to deliver breathable air, drinkable water, nutritious food, waste disposal services, a sense of place, biological diversity, cultural identity and spiritual sustenance. So our social and economic planning should explicitly take into account the limits of natural systems, rather than assuming we will always be able to clean up any mess we make as long as the economy is booming.

It will have a Long Time Horizon, recognising that the decisions we make about population, urban planning, transport facilities and energy supply will have long-term consequences. Indeed, I have argued that the choices we make in these areas literally set in concrete the options for future generations. It should be a matter of concern that these choices are often dictated by this year's balance sheet or next year's election. We would make better decisions quite generally if we routinely asked ourselves what they would look like in 50–100 years' time. In some cases, that level of foreknowledge is not possible, but we should at least be making the effort to invest in developing a greater awareness of the long-term consequences of our choices.

The sustainable future will be Informed, in the sense of better understanding the Earth's complex natural systems. We will be able to make more responsible choices if we are more aware of the consequences. Much of the damage we have done to natural systems has been caused by ignorance rather than wilful vandalism. We urgently need to invest more resources in order to understand the complex natural systems and the toll of human activity. We also need to use that informed perspective to make the hard decisions needed; those choices will only be politically sustainable if the community is involved through a participatory process, such as that used by Douglas Shire in north Queensland to decide its method of water treatment. The council commissioned engineering studies of four alternatives, costed to estimate the impact on local rates, and supported 18 months of public debate before a plebiscite. The community voted strongly to use the cleanest, but most expensive, option.

It will be Efficient in the sense of turning resources much more effectively into the services that people want. A range of studies show that efficiency improvements of a factor of four are achievable with technology that exists and is economically viable today, while factors of ten are reasonable 'stretch targets' in most areas. So we should be able to reduce by a factor of five to ten the resources needed to produce essential services such as warmth, shelter, clothing and food. This is an urgent political priority because people in the poorest parts of the world will only be able to enjoy the material comforts we take for granted, such as decent shelter, clean drinking water and adequate nutrition if these services are provided more efficiently. It would take several planets to allow everyone to live as wastefully as we now do. To demonstrate the extent to which this is a political problem rather than a physical limit, as mentioned earlier, the UNDP has calculated that all humans could be given those basic building blocks of a civilised life, as well as basic health care and education, for about 5 per cent of the global military budget! Rather than spending more and more obscene amounts preparing to fight each other for the dwindling resources, we should be investing in the technologies that will allow all to live comfortably within the limits of the Earth's resources and capacity to process our wastes.

Finally, the sustainable future will be Resourced because we will have made the transition from fossil fuels that are either limited, like oil, or unacceptably polluting, like coal, to the abundant flows of renewable energy. The crucial point is that the flows of renewable energy are huge by comparison with any conceivable future human demand. The Department of Resources and Energy report I mentioned earlier estimated that we could develop the storage systems that would allow us to get all our electricity from renewable energy supply systems by 2030. It would be more expensive than the present supply mix, mainly based on coal, but it would not be changing the global climate. The technical challenge of powering Australia from a mix of renewable energy supplies by 2030 is much less demanding than the goal President John F Kennedy set the US in 1960 when he said they would try to get a human to the Moon and back within the decade. Like that goal, however, it will only be achieved if there is real political will.

The Earth Charter Initiative is a broad-based, voluntary, diverse global network of people, organisations and institutions, which participates in promoting and implementing the values and principles of the Earth Charter. Members include leading international institutions, national governments and their agencies, university associations, non-government organisations and community-based groups, city governments, faith groups, schools and businesses – as well as thousands of individuals. The Earth Charter embodies the fundamental principles of a new global ethics under four broad headings: 1) respect and care for the community of life; 2) ecological integrity; 3) social and economic justice; and 4) democracy, non-violence and peace. Australian environmental groups have worked strongly for decades on the first two of these. There is now an increasing recognition that solving the complex problems we face demands an integrated approach. Solutions to our environmental problems will not be socially and politically sustainable unless they are based on social and economic justice. This is most obviously true of climate change, which can only be solved if the developed nations accept their responsibility for causing the problem as well as accepting the legitimate demands of poorer countries to improve their material wellbeing. These principles lead inexorably to a 'contract and converge' model in which the Earth's capacity to absorb carbon dioxide is much more equitably allocated.

I have also argued that we will only achieve durable solutions to the complex problems we face if we develop participative structures that involve the whole community in making the difficult decisions. In the modern world, the old 'decide and defend' model of government decision making inevitably spawns motivated and well-organised pressure groups determined to prevent the government getting its way. So the scale and complexity of the problems we face is demanding a different approach, one based on the new values which are the only credible foundation for a future sustainable society. In many ways, our old-fashioned institutions are impeding our progress toward this goal. Our universities and research institutions could be argued to reflect successive waves of intellectual fashion, laying down in layers for future intellectual stratigraphers the evidence of our inability to develop transdisciplinary approaches to the complex problems we face. So there is a clear role for independent scholars, free of those institutional fetters and therefore able to develop the new thinking and different approaches we desperately need.

I have extended the metaphor used by the US economist Lester Thurow when he said that it is difficult to tell people the party is over when they haven't yet got to the bar! In that sense, I argued, we have a responsibility not just to dash the hopes of young people who understandably aspire to living at least as well as their parents have. We have to tell them that the old party is over because it was based on a series of delusions: that resources are unlimited; that we can grow forever on a finite planet; that the Earth will rebound from any assaults we mount on its natural systems. But we can also tell them about a new party that is starting up, to which they are warmly invited. It will be a better party for several reasons. It won't run out of food and drink because it will be based on sustainable use of the Earth's resources rather than a loot, rape and pillage approach. It will be more rewarding because it will be based on fulfilling personal experiences rather than gluttonous consumption. It won't leave us with a nasty hangover of radioactive waste or degraded landscapes. It won't have the neighbours enviously peering in or throwing rocks on the roof, because they will all be invited. Most fundamentally, it is a party that we can be confident future generations will be able to enjoy. I hope we are able to work together to give our descendants that sort of legacy.

AUSTRALIA'S ENVIRONMENTAL CRISIS

Excerpt from a paper prepared for the National Civil Society Dialogue, August 2006

We have a beautiful and unique environment and many aspects of it are in good condition by international standards. But several national reports have documented the scale and seriousness of environmental problems: loss of biological diversity; degradation of inland waterways; and destruction of the productive capacity of rural land. These problems are getting worse, because the pressures on natural systems are still increasing. Each year there are more of us and we use more energy, travel further in larger and less efficient cars, live in larger houses, consume more resources and produce more waste.

The decline is confirmed by Australian Bureau of Statistics reports on measures of Australia's progress. Since 1990, all of the usual economic indicators show positive trends, but almost all the environmental indicators are getting worse. We are funding unsustainable levels of material consumption by running down our natural capital. Or, to put it in economic terms, we are operating our ecological accounts at a heavy deficit for which our children will pay.

The evidence that our current way of life is unsustainable is overwhelming:

- We are overusing water and degrading our major river systems. Water restrictions are now semipermanent.
- We are seriously changing the global climate, with economic and social consequences ranging from increased costs of water supply to growing numbers of human casualties from heat stress and severe weather events.
- We are already in the middle of the sixth major extinction event in the history of the planet, with global warming adding to the driving forces of habitat loss, introduced species and chemical pollution.

Many of Australia's natural treasures are under threat:

- The Great Barrier Reef – arguably our most outstanding natural asset – faces a range of pressures, including shipping accidents and oil spills, run-off from agricultural activity on the mainland, trawling and overfishing, as well as coral death or 'bleaching' from warming waters.
- Australia's greatest river system – the Murray Darling – is crucial to the environmental and economic health of New South Wales, Victoria and South Australia, yet is suffering from extreme overextraction that is leading to decline in ecosystem health and water quality. In addition, increasing evaporation and reductions in rainfall resulting from climate change is expected to lead to a 50 per cent decrease in water flowing into the Murray Darling system.
- Our unique plants and animals – Australia has an unusually high proportion of species that are not found anywhere else, which means that a species lost from Australia is lost to the entire world. Land clearing, climate change and other threatening processes are placing Australia's unique biodiversity at great risk.

- The Kakadu wetlands are at risk from climate change, with up to 90 per cent expected to be lost with a 2–3 degree rise in average temperatures.

So what can we do to achieve a sustainable future? We need to move beyond the simplistic view that economic growth will solve our problems. We need a different approach, one that recognises our responsibility to future generations. There is a growing awareness around the world that a sustainable future will involve significant change.

Achieving a transition to a sustainable future will require fundamental changes to our technologies, our social institutions and our values. Human systems can change radically and quickly and the transition we need may be catalysed by growing community awareness of the problem. Unfortunately, our decision makers and opinion formers still behave as if there is no problem, or see potential solutions as threatening their short-term interests.

We should see the economy as a means of serving our needs within the limits of natural systems, rather than an end in itself. We need a technological transition based on the principles of renewable resources, efficient use and 'industrial ecology' – using the waste of one industrial process as the feedstock of another. We can eliminate hunger if we stabilise our population and improve distribution systems; the world now produces 2 kilograms of food per person per day, more than enough if it is equitably distributed. Above all, we can create a future of genuine globalisation, recognising that we share a common fate with the whole human family, rather than the false globalisation that considers only economic issues.

There will always be some who say we can't afford to do things better. As the International Chemical Secretariat showed in its recent report, *Cry Wolf*, some vested interests have always resisted change by overstating the costs and ignoring the benefits. When the catalytic converters that have dramatically cleaned up our urban air were proposed, some in the car industry claimed they would cost over $1000 each with a fuel consumption penalty on top, for no obvious benefit. In fact, they cost about $100 each, led to more sophisticated engines and improved fuel efficiency, and are estimated to have reduced healthcare costs in the UK alone by about $5 billion a year. It was claimed that measures to clean up sulphur dioxide from power stations and stop acid rain would add 25–30 per cent to electricity costs; they had no significant impact on prices. When regulations to clean up coal mining were proposed in the US, industry claimed it would cost between $6 and $12 per ton; it cost less than $1. The Australian Business Roundtable on Climate Change has concluded that we can afford to take strong action to reduce greenhouse pollution. More importantly, strong action now will be much better for the economy than inaction now, leading to a need for much more drastic measures in the future.

New technology and improved efficiency are crucial, but they won't achieve a sustainable future unless we also embrace new values. Rather than the inevitably futile path of trying to dominate nature, we need to understand the limits of natural systems and live within those limits. Rather than continuing to erode the social fabric for short-term political gain, we must develop social institutions that will allow us to work together to solve our difficult problems and take the hard decisions needed for a sustainable future. Rather than seeing the level of material consumption as an end in itself, we should recognise that consumption is, at best, only a means to the end of greater satisfaction.

Clearly existing policies and institutions are failing. A national Sustainability Charter and Commission would provide a coherent national and long-term framework to drive investment, policies and programs to deliver a sustainable Australia within a generation.

Enshrined in the charter and delivered through an independent and effectively resourced Sustainability Commission, ecological parameters would be set to guide investment, policies and programs across the economy. The proposed Sustainability Commission is a similar concept to the Australian Competition and Consumer Commission and would report to COAG. It would provide the principles and parameters for state implementation, supported by local engagement.

NOW OR NEVER

A response to Tim Flannery's Quarterly Essay 31: Now or Never, *published in* Quarterly Essay 32, *Black Inc., 2009*

Tim Flannery is right to say that we face a planetary emergency. Climate change is not just a serious threat in itself, but also a factor compounding other major environmental problems like the catastrophic loss of biodiversity, increasing water shortages, the degradation of rural land and the decline of fisheries. Four reports in the UNEP series on Global Environmental Outlook, the Millennium Assessment and the report of the International Geosphere-Biosphere Program have all reached the same conclusion. As *GEO2000* said, the present approach is not sustainable and postponing action is no longer an option. It has been clear for 20 years that the rate of burning fossil fuels poses the risk of dangerous interference to the global climate system. Our current situation has been made unnecessarily difficult by decades of government inaction, largely driven by a naive faith in market forces and, until last year, our national government's studied refusal to accept what the science was clearly saying.

Meeting our global responsibility for climate change demands rapid reduction in carbon dioxide emissions. The IPCC's fourth report, released last year, urged the big polluters like us to reduce our emissions 25–40 per cent by 2020, as part of a reduction of at least 80 per cent by 2050. More recent science suggests that these targets may not be enough to prevent dangerous climate change. A cautious approach that recognises the uncertainty of the science would lead to greater reductions. Some are suggesting we should aim for a 50 per cent cut by 2020 and 95 per cent – effectively decarbonisation – by 2050. Any serious targets will require attention to electricity supply and use, transport and agriculture. The other big problem looming, 'peak oil', compounds the difficulty of shaping a coherent policy. As with climate change, we have known for decades about the problem, but decision makers have been in denial. Some alternative transport fuels that looked promising, such as shale oil, tar sands and liquid fuels from coal, are now effectively ruled out by the need to reduce our release of carbon dioxide.

So what should be the basis of our response? We need the sort of serious targets specified above and a comprehensive approach to the whole issue. As the Prime Minister said in his closing speech to the 2020 Summit: 'Climate change is the overarching issue this generation and those to follow must address.' He concluded that it should be considered in all decisions. I agree. It should, for example, have informed the recent response to turmoil in financial markets. As Tim Flannery correctly observed: 'While a high price for carbon is absolutely necessary, it alone will not be sufficient.' Recent events should have demonstrated beyond any doubt that it is irresponsible to leave an issue as important as the survival of civilisation to the capricious behaviour of markets. As well as a realistic price for carbon, certainly at least $50 per tonne, we also need serious targets for the share of our electricity from renewable sources, for appliance efficiency, for the performance of domestic and commercial buildings. We should also pay attention to urban planning, ensuring that the services people use often are within easy reach on foot or bicycle, as well as investing in a massive upgrade of public transport for longer journeys.

So far, I agree completely with Tim Flannery. Where I part company is on the issue of 'clean coal', the new technological saviour now that nuclear power's feet of radioactive clay have been exposed. I just don't think it is realistic to put all our eggs in that very fragile basket. It is certainly true that we have to do something about the overwhelming dependence on coal for electricity in this country. The biggest single contribution to climate change is carbon dioxide from burning coal, mainly to generate electricity. Coal is a very dirty fuel for two fundamental reasons. Because it always contains a range of impurities, burning it produces a cocktail of gases, like oxides of sulphur and nitrogen, as well as soot, ash and heavy metals, even some radioactivity! More basically, coal is mostly carbon, so burning coal gives carbon dioxide, the most important greenhouse gas. Oil and natural gas contain significant amounts of hydrogen, which burns to produce water. As a result, those fuels produce less carbon dioxide per unit of energy. Brown coal has a third problem: as it is chiefly water, most of the energy produced by burning is used to evaporate the water. It thus gives even less useful energy for every kilogram of carbon dioxide released.

So how could burning this dirty fuel be cleaned up? Some believe we could capture the carbon dioxide from burning coal, compress it into a liquid and inject that into rock layers deep underground. This does nothing to reduce the other forms of pollution, so talk of 'clean coal' is the height of chutzpah. Because air is nearly 80 per cent nitrogen and only about 20 per cent oxygen, carbon dioxide is only a small fraction of the exhaust gases from a power station. Separating it from the other gases is complex, expensive and takes large amounts of energy. Some propose burning coal in pure oxygen rather than air, so the exhaust stream would be mainly carbon dioxide. That would simplify separating the carbon dioxide, but it would be expensive and energy-intensive to produce enough oxygen to burn bulk quantities of coal. Compressing the gas into a liquid is technically feasible, but it also takes large amounts of energy. The liquid carbon dioxide would then have to be transported to sites where the rocks are judged suitable for storage. The World Energy Council has estimated that the carbon dioxide from the world's power stations would, if compressed to liquid for transport, be a volume comparable with the entire global oil and gas industry! While there are some sites that look suitable for long-term storage, it has not yet been shown that they would be secure for geological time; obviously, we could not afford to have the gas leak out into the air. There may be some sites secure enough and close enough to power stations for a few demonstration projects, but it is pure moonshine to believe this approach could be applied to all future power stations.

Even if all the technical problems could be solved, the extra energy required for the process would mean that carbon dioxide released per unit of energy would only be reduced about 70 per cent. We need to go much further than this by 2050 to stabilise the global climate, so it makes much more sense to invest in the really clean, renewable forms of energy like wind, solar and geothermal. There are no credible estimates of the possible cost of carbon capture and storage, but industry sources talk openly about it possibly adding 50 per cent to the price of coal-fired power. Tim Flannery suggests that a carbon price as high as $80 per tonne might be needed for 'clean coal' to be economic. Studies of the prospective renewable supply systems show they could achieve greater reductions at lower cost – and much faster. Even if everything were to work as well as the optimists hope, it would be many years before the technology could be credible. The most that could be claimed is that 'clean coal' might perhaps slow the growth of greenhouse pollution after 2020. Climate science is now saying we face an emergency and need to cut emissions significantly before then. While the whole idea is still unproven, even if it could be made to work it would do too little, too slowly and at too high a cost. So why is it being taken seriously?

There is no simple answer. Coal companies want to stay in business, so they have an obvious commercial reason to claim that we can keep burning coal. At the 2020 Summit earlier this year, I said that there are differences of opinion about the feasibility of 'clean coal', but we could surely all now agree that it is irresponsible to continue 'dirty coal' by building old-fashioned coal-fired power stations. A few delegates, mainly from the industry, objected – so Senator Penny Wong had to report that there was no consensus for the position I advocated. I was shocked to find that there are still people in the coal industry who think it is acceptable to plan projects that will belch out millions of tonnes of carbon dioxide. As well as business interests, some unionists representing coal workers – and some ALP politicians close to those unions – support 'clean coal'. The companies think it would be acceptable to keep the Earth habitable if it doesn't reduce their profits; the labour interests think it would be acceptable if no jobs are lost. To be fair, jobs are at stake. The coal industry, domestic and export, employs more than 25,000 people. To put that figure into perspective, while John Howard was prime minister 150,000 jobs were lost from manufacturing. In other words, the average annual loss of manufacturing jobs was about half the total employment in the coal industry. Those workers were absorbed in other fields of employment. Expanding clean energy supply and efficient use will create far more jobs than the coal industry now provides. Training and other forms of assistance would help the workforce move into the more satisfying jobs that would be created.

Some technocrats still hope for a big technical fix to the problem of climate change. So they are attracted to grand technical delusions like 'clean coal' or nuclear power, rather than accepting the wisdom of applying simpler technologies like wind turbines, solar cells or improved efficiency.

The final complexity is the level of political support, including public endorsements and huge allocations of funds. Even politicians who accept that climate change is a real and urgent problem are reluctant to propose the changes we need. They cling to the old myths: business can go on as usual; that further growth is tolerable; even that growth should be encouraged. We are now seeing a concerted campaign by some industry interests to dilute still further the Garnaut report's recommendations for an emissions trading scheme. Garnaut has supported a very gentle start to the scheme, with prices as low as $20 per tonne of carbon, but vested interests are clamouring to be sheltered even from this charge. A report released by the Australian Conservation Foundation calculated that the proposed public support of 'trade-exposed industries' would cost the taxpayer billions, most of which would be handouts to very profitable overseas-owned corporations which employ relatively few Australians. This is wasting money we should be using to stimulate the new industries needed to meet our reduction targets. As Tim Flannery points out, a carbon price of $40 per tonne would probably make it commercially viable to produce from plants a fuel called biochar. Various studies have concluded that a range of renewable supply technologies would be economic at $50 per tonne. We should face up to the need to completely re-invent our electricity system over the next 40 years so it will have been decarbonised by 2050. That will require both a realistic price for carbon emissions and a range of other government support measures.

I criticised the Howard Government for pursuing the ridiculous distraction of nuclear power, rather than responding to climate change. The equally simplistic notion of 'clean coal' is now playing a similar role for the Rudd Government and state governments along the eastern seaboard: it is a simulated response, aimed at appeasing the public concern about climate change without antagonising business, unions or those troglodytes in parliament who still can't see the problem. This is the real danger of the 'clean coal' bandwagon. It is frittering away the time and resources that we should be using to restructure our energy supply and use. The Rudd Government should trust the community and engage us all in developing a concerted response strategy, as was done in Sweden. We need that sort of participative process to produce a politically sustainable strategy to slow climate change. Unless we achieve that, we all face a very bleak future.

As a final point, the fundamental message of Tim Flannery's essay and the various reports I cited is that we need to recognise the limits of ecological systems and build that recognition into our planning.

Allowing markets to decide how resources will be used and how waste products will be managed is ethically indefensible. It means that the selfish wishes of today's affluent minorities take precedence over the needs of all future generations. We also need to get over our universal fixation on growth. There is no prospect of a sustainable future unless we stabilise the human population and our per capita consumption at levels within the capacity of the biosphere. As a concrete example, I agree with Tim Flannery that we should utilise geothermal energy, but I would not support the idea of a massive mineral processing industry in the Cooper Basin. We already face the prospect of a lunar landscape stretching for 100 kilometres or more if the bizarre plan to expand the Olympic Dam mine goes ahead. In a carbon-constrained world, it makes no sense to base our economic future on exporting huge volumes of low-value commodities. We should instead be investing in the growth industries of the 21st century like clean energy supply, efficient energy use and water-efficient food production.

ECONOMICS AND POLITICS

INTRODUCTION: THE INEXTRICABLE MIXING OF ECONOMICS AND POLITICS

Not long after I came to Queensland, the local newspaper conducted a survey of political knowledge among undergraduate students. There were such basic questions as the name of the prime minister, the name of the state premier, the name of the lord mayor of Brisbane; more demanding questions asked which level of government was responsible for various services, roughly how many members sat in different houses of parliament and so on. The general level of knowledge was unimpressive. The lowest score, two out of ten, was achieved by a student who giggled and said, 'I don't know anything about politics, I'm studying economics.' I wondered at the time how it could be possible to teach economics and give a student the impression that it had nothing to do with politics.

I think part of the explanation lies in a particular emphasis within the discipline of economics. A colleague claimed in the 1970s that economists had 'physics envy'; a desire to wrap their subject in layers of complex mathematics to give the impression that it has the same level of objectivity and reliability as scientific knowledge. We had both witnessed a presentation by an economist who covered the board with mathematical functions and differential equations to justify his assertion that a perfectly free market will utilise a natural resource at a rate that is optimal, not just for this generation but for all future generations! The discussion centred around an obvious point. If the rate of utilisation of a resource like oil is optimal for this generation, it is almost certainly not going to appear optimal – or even defensible – to people who live in a hundred years' time.

Most mathematical expressions of economics are based on premises that are clearly invalid: perfectly free markets, with perfectly informed individual actors making decisions based solely on a desire to optimise their economic outcomes. Elementary logic tells us that if you start from false premises and argue logically you must reach false conclusions. That style of economics can only produce valid outcomes if its logic is also flawed. There is no such thing as a perfect market, there never has been and there never will be. Government policies always allow some activities, encourage some, discourage others and try to prohibit others. There is clearly a demand for illegal substances, which governments actively discourage, while government subsidies encourage other activities; for example, at the time of writing we were providing generous subsidies to the aluminium industry, allowing it to operate profitably using coal-fired electricity in competition with Icelandic geothermal energy and Canadian hydro-electricity. There was a brisk debate at the time I wrote this introduction about the huge subsidies of road freight and the consequent risk to other road users who share the road with large trucks.

In the real world, political decisions shape our economy and the state of the economy shapes our politics. One observer has suggested that politics in the modern world is almost synonymous with economic management. Most politicians believe that they are unlikely to be re-elected if the economy is going badly and unlikely to lose power if the economy is in good shape. The writings I have collected in this section are intended to illustrate the complex intersection of economics and politics: how political decisions are shaping our economy and how economic choices are affecting local politics.

OUR ECONOMIC FUTURE

Extract from A Big Fix, *Black Inc., first edition 2005, revised edition 2009*

Our economic future depends absolutely on the capacity of the natural world to provide food, fibre and minerals for our own use and for export. It also relies on the capacity of the natural world to process our waste. These are fundamental issues. Equity and social stability are also essential aspects of our economic future. As far back as 1776, Adam Smith argued that economic activity requires what he called the 'social bond'. Without social stability, he argued, people wouldn't lend money, invest in new ventures or even work for a living, since all these activities rely on the presumption that agreements will be honoured. An increasing concern about the extreme market approach of right-wing economists (and most of our governments) is that it inevitably widens the gap between rich and poor. Again, this is not a new flash of insight; the 19th-century economist Vilfredo Pareto reached the conclusion that a market approach would inevitably lead to a smaller and smaller group possessing more and more of the wealth. Inequality inevitably erodes the social stability that is needed for the orderly operation of markets, so the extreme market approach is manifestly self-defeating.

Australia's enduring economic problem is that we have the trade pattern of a Third World country. We pay for our imports of goods and services by exporting minerals and farm produce. There are two reasons this will inevitably lead to economic decline. The first is that due to technological advances, smaller and smaller amounts of raw materials are turned into increasingly expensive end-products by the application of human ingenuity. A typical laptop computer selling for $5000 contains about $50 worth of raw materials and the remaining $4950 is payment for the human organisation of those into a finished product delivered for our use. With the exception of resources that are in limited supply, such as oil, the relative value of raw materials is steadily decreasing and the relative value of manufactured goods and elaborate services is steadily increasing. So an economic strategy of selling raw materials to pay for value-added products means you have to sell more and more each year to cover the costs of the imports. The problem is compounded by a growing population and increasing consumption per person, which means that the volume of material exports needed to pay for our imports is rapidly increasing. At the time of writing, Australia had just recorded a trade deficit for the 43rd consecutive month. That looks like a serious trend to me.

The second reason is more fundamental. If you earn your living by taking raw materials and converting them into value-added products, you can always increase production to fulfil growing needs. Since we earn our living by selling our minerals, we can't increase production forever; production is limited by the finite nature of the natural resources. We have tried to produce more than the sustainable yield from our agricultural land to meet the costs of our imports, but we are paying the price in degradation of rural land. Essentially the Australian economy has produced the illusion of success by liquidating our natural capital, like a farmer who sells a piece of the farm each year to counterbalance an operating loss. Successive governments have been running the equivalent of a Ponzi scheme, a particular type of fraud for which you can be prosecuted. The essence of a Ponzi scheme is that shareholders in a company are paid dividends out of the capital assets rather than profits. In the short term, all appears well and the investors receive their dividends, but at the cost of the viability of the operation in the longer term. As shareholders in Australia Incorporated, we have been paid generous dividends over the last few decades by running down our natural capital. Every tonne of copper, coal or iron exported from Australia; every cubic metre of gas sold to overseas customers; every kilogram of beef produced by degrading rangeland represents the sale of another chunk of our natural capital, another piece of the national 'farm' gone forever – and another step toward a future of collective poverty. If the natural capital were turned into social capital, for example by investing in useful infrastructure, the transaction could have enduring value. But most of the profits from the sale of our mineral assets go to the overseas interests that own the companies concerned, or are frittered away in consumption that provides no enduring benefit.

The liquidation of our natural assets to boost the Australian economy has been supplemented in recent years by growing tourist income. Overseas tourists tip money into the local economy and provide thousands of jobs. The figures for particular local areas are impressive and show the value of using natural resources in a way that might be sustainable, rather than destroying them. Tourism associated with the Ningaloo Reef brings about $120 million a year into the West Australian economy. Tourist activity linked to the wet tropical rainforests turns over hundreds of millions of dollars a year, compared with the paltry $13 million produced each year by logging those forests before it was halted. Great Barrier Reef tourism is estimated to yield about $3 billion a year. The dramatic recent increase in tourism has been driven by relatively cheap international air travel, which is likely to be a casualty of rising oil prices. Tourism is also critically dependent on perceptions of personal safety, as the resorts of Bali and Thailand have recently discovered. Tourist income is a useful supplement, and certainly a more rational way to exploit our unique natural resources, but we should acknowledge the inherent fragility of the income stream. We should also recognise that most tourists come here because of our natural assets, so maintaining them is crucial to this area of economic activity.

A durable future economy will have two characteristics. It will use the income stream from natural resources, rather than rapidly depleting the stock, and it will use human ingenuity to transform these assets into useful goods and services that people want to buy. Some state governments recognise this. Former Queensland Premier Peter Beattie tried to change his region from the Sunshine State to the Smart State, since he could see that there is no economic future in continuing to sell farm produce and minerals. South Australia and Victoria have also made significant investments in science and innovation. Successive Commonwealth governments have, however, systematically eroded our capacity for science and innovation. CSIRO, once internationally renowned for the quality of its research, has seen its funding steadily cut back and been forced to rely more and more on short-term consulting. University research has been starved of funds and specialised granting programs like the national energy research scheme have been abolished. When the Opposition proposed a 'knowledge economy' at the 2001 election, the Coalition government and the commercial media derided the idea. Public assets have been sold off to private interests, in many cases not even based in Australia, thus steadily reducing our capacity to make decisions in our own long-term economic interests. I find it hard to imagine how any government could support the sale of public utilities such as the suppliers of water, communications and power. Young kids know you don't sell the utilities when you are playing Monopoly, because they provide a secure income stream! The sale doesn't even make sense within the narrow framework of market economics; as Adam Smith pointed out in 1776, monopoly control of an essential service will always lead to profiteering and sub-optimal provision of services. Major city airports are classic examples; sold to private enterprise, they now charge more and provide worse services to the travelling public to ensure profitable returns to the owners.

The entire notion of economic planning has been abandoned in favour of a naive faith in the magic of the market. The market can respond to short-term fluctuations in demand and supply, but there are obvious problems when the time lag is long between perceiving a shortage and being able to respond. Landholders in the late 20th century saw the cultivation of grapes and olives as profitable, so the areas of land devoted to growing these crops have significantly expanded. When all enter into full production, there could well be an oversupply that will depress prices, casting doubt on the investments. Investors saw a growing demand for inner-city apartments, but the oversupply led to declining prices and questionable economic viability for many of the ventures. Quite generally, the market won't provide the long-term investments needed unless there is an opportunity for large profits to recoup the money outlaid. Trusting the market means accepting the cycles of boom-and-bust that are an inevitable feature of a market economy.

The principles of a more rational economics were laid out over 30 years ago by Herman Daly, who has been an economics professor in various US universities as well as a senior advisor in the World Bank. Daly argued for a three-stage process. First, he said, we need to ensure that the total scale of human activity is ecologically sustainable. Second, we should distribute resources and property rights fairly, both within this generation and between generations, also taking into account the needs of other species. Finally, Daly proposed, we should try to allocate resources as efficiently as possible within these constraints. So there is a role for markets in ensuring efficient allocation of resources, but first, science must determine the scale of resource allocation we can responsibly allow. Society then needs to work out the principles of fairness within which markets can operate. This is a much more sophisticated approach than taken by most economists, who Daly believes 'chase an assumption of infinite wants along the road of infinite growth'. If we know that the present level of resource use is straining the capacity of natural systems, encouraging further use of these resources is totally irresponsible.

We must transform our unsustainable economy into a sustainable one. There can surely be no argument against this goal. In pursuing this aim, we should be as flexible and adaptive as possible.

NOT ALL HAVE A VOICE IN THE FREE MARKET

The Weekend Australian, 27–28 April 1991

Two hearty cheers for the mythical free market! It doesn't exist, but it might be a good thing if it did. Even though real markets don't live up to the romantic ideal, they allocate resources in an economically efficient way. In a market economy, the demand for goods and services represents the sum total of consumer choices. As a result, markets respond to changing demand patterns much more rapidly and flexibly than systems of central control.

Unfortunately, there is a tendency to see the market mechanism as the answer to all the problems of resource allocation and consumer behaviour. This approach has a fundamental flaw. Many aspects of consumer behaviour are not driven by economics. We all make a range of decisions which arise from our social situation rather than an economic analysis of the best use of our money.

Additionally, a typical consumer rarely has the information which would be needed to make economically rational decisions. Electricity users are typically unable to determine the rate of using power or what it is costing them. The situation is complicated by the advertising industry, which could be said to exist mainly for the express purpose of ensuring that consumers do *not* make economically rational choices. We are regularly encouraged to believe that we will be out of fashion or a social outcast unless we waste our money on some trendy item.

In recent discussions about patterns of energy use which could be ecologically sustainable, I found that many people wanted to leave decisions about energy use to the market. If we want the market to encourage a move toward sustainable development, we have to ensure that the price signals give consumers the right message. There are some basic problems in doing this.

The prices we pay for fuels such as petrol, gas or electricity at the moment do not include several important factors. The prices do not include the costs of long-term environmental damage, make no allowance for resource depletion, take no account of the social impacts of fuel use and ignore the macro-economic consequences of our choices. Thus fuel prices do not compensate future generations for the depletion of resources, do not include the costs of acid rain or climate change, and do not account for the impacts of fuel use on patterns of social life. Prices make no provision for what fuel purchases do to our balance of trade or the strength of the Australian dollar. More importantly, there is little prospect of developing formulae that would make it possible to build these factors into the price structure. Market prices simply do not reflect the wider considerations which determine whether our resource use can be sustained, so market choices also neglect those wider issues.

There are some possible policy approaches that could be used to modify the price signals. We could devise mechanisms, such as a carbon tax, which would alter the price signals and affect the choices made by consumers. But in the absence of some intellectually defensible means of fixing the scale of such taxes, we can't be sure whether they will even come close to reflecting the real cost of our resource use.

There are some other problems with the market approach. One

of the key elements of sustainable development is equity between generations, ensuring that our resource use does not unreasonably close off options for our descendants. Since only the current generation can participate in the market, the wishes of future generations of consumers cannot possibly be reflected in the market choices. Even if prices do not explicitly discount the future, the needs of future Australians will always be discounted by their inability to influence the current market.

The inequalities in market power between individuals, groups, regions and nations are well-known problems. A doubling of the petrol price might have a severe impact on the budget of a typical household in the outer suburbs, but have no effect at all on the travel of a business executive or senior public servant whose fuel costs are paid as part of a salary package.

There are some more fundamental difficulties if we leave decisions about resource use to the market. Sustainability requires the conservation of biodiversity and ecological integrity. It is difficult to see how these issues can be left to a market approach. Tasmanian economist David Heatley has suggested that we need to recognise that the human economy coexists with a range of 'natural' economies. He argues that the 'possum economy', consisting of food, trees and nesting sites, runs according to laws of supply and demand like our economy. There is trading between the possum economy and the human economy. My mango tree is a positive contribution to the possums, providing an extra source of food; other human activities, such as clearing bush, impact negatively. Unless they run out of the bush and attack us in such a way as to change our behaviour, possums cannot influence the human economy. Our economy does not take account of the loss of biodiversity or damage to the integrity of natural ecosystems. We can load the prices of those activities we recognise to be damaging, but the loading will inevitably be arbitrary.

Appropriate pricing is clearly one tool to influence consumer behaviour in socially desirable directions, assuming we can agree what those are. We can load on to the current fuel prices, which broadly reflect supply costs, additional components to reflect resource depletion, environmental damage or even impacts on natural systems. We need, however, to recognise that those loadings will always be arbitrary, and so will reflect value choices about the extent to which we value diversity, natural systems or resource depletion. Unless we recognise those factors, leaving decisions about resource use to the market is almost criminal negligence.

THE CLUB OF ROME'S GOOD OIL

The Weekend Australian, 11–12 August 1990

The outbreak of hostilities in the Middle East has drawn attention once more to the fragility of the world's supply of oil. Totally dependent on massive amounts of oil for transport fuels, most industrial countries grow anxious when the political stability of the oil-rich region is threatened.

It's an ill wind that blows nobody good, however. The oil embargo in 1973 certainly stimulated exploration of better ways of fuelling industrial societies. A newspaper column earlier this week caught my attention with its reference to the first oil shock of 1973, the second in 1979 and the subsequent decline in oil prices. This, it was said, was 'predicted by those economists who had already dismissed the alarmist and simple-minded scenarios of the Club of Rome'.

I have been intrigued before by this tendency to portray the Club of Rome as a group of simple-minded folk who didn't understand the real world. Even though the members were leading scientists, industrialists and public servants, it has been suggested that their ideas should not be taken seriously. Since few of them had achieved the intellectual heights of undergraduate studies in economics, they inevitably spoke from a position of comprehensive ignorance. The conclusions of their 1972 report, *Limits to Growth*, have been depicted by generations of economists as 'alarmist and simple-minded'.

It is not clear how thoroughly those people have read the report of the Club of Rome. I am sure that this country would be in much better shape if even half the people who misuse Donald Horne's ironic phrase 'the lucky country' had actually read his book; but that's another story. The computer model developed for the Club of Rome was the first crude attempt at global modelling of long-term trends. It had several shortcomings but still yielded some useful insights. Let me remind you of the three main conclusions.

The first was that if the existing trends in population, food production, industrial output, resource use and pollution were to continue unchallenged, limits to growth would be reached within a century. It is hard to believe that this was seen as contentious. We are already seeing global environmental problems such as the greenhouse effect from a population of 6 billion people and the present level of industrial development. Since the human population is doubling around every 35 years, another century of growth at the current rate would involve about an eight-fold increase in the human population. At the same time, industrial development is a goal of most countries and further growth is almost a universal ambition. Could anyone seriously argue that a future world population of about 50 billion could be sustained, at a higher level of industrial development, without reaching environmental limits?

The report's second main conclusion was that it is possible to alter present trends and establish conditions of economic and ecological stability that could be sustained indefinitely. If that is alarmist talk or unwarranted pessimism, may we be surrounded by such pessimists! It might be fair to accuse the Club of Rome of starry-eyed optimism, or unreasonable faith in the ability of our leaders to see beyond short-term economic considerations to the conditions for sustainability, but they can scarcely be accused of spreading gloom and doom.

Their message was that we have a choice: we can either stabilise the human population, levels of resource use and the pollution we produce, or we can carry on at the risk of overtaxing the natural systems of the Earth. Since such problems as the greenhouse effect and the depletion of the ozone layer imply that we are already stretching our luck, the warning almost 20 years ago was timely and perceptive.

The third conclusion was even less contentious. If the people of the world wish to achieve sustainable development rather than pursuing the delusion that present levels of growth can continue indefinitely, the chances of achieving that goal will be improved by starting sooner rather than later. That is self-evident: the longer we delay, the more serious are the problems. It is true of the greenhouse effect, the ozone 'hole', preserving agricultural land, husbanding non-renewable resources and so on.

It is not hard to see why the conclusions of the Club of Rome were so threatening. One obvious reason is that such thoughts pose a clear challenge to the naive belief that the market will solve all the problems of resource depletion and accumulated pollution. Now there's a simple-minded view!

The argument is that if people really want more resources or reduced levels of pollution, the market will deliver those requirements. One problem of this approach is the lead time for markets to respond to new demands. If you were in a bus that went over a cliff, it would not be wise to assume that the sudden market demand for parachutes would produce them before you hit the ground! In the case of fuel, it will only make economic sense to invest in an alternative such as alcohol when the price of oil is high enough for the alternative to be profitable. But it would probably take ten or 20 years to increase production to the point of replacing a large fraction of our oil. Thus leaving such investment decisions to the market is a guarantee of decades of deprivation or dependence.

Far from being alarmist and simple-minded, the Club of Rome drew the world's attention to important problems, pointing out that continued growth would eventually tax the natural systems of the Earth. They did say we had time to plan a transition to sustainable living, but we have frittered away two decades of that time. Concerted action is long overdue.

PRIVATE BEATS PUBLIC? IT'S A MYTH

The Weekend Australian, 16–17 June 1990

At the time this column was written, we had two domestic airlines, a privately owned one which later collapsed – Ansett – and a publicly owned one which was later taken over by Qantas. What is now Telstra was called Telecom, and Telecom and the Commonwealth Bank were publicly owned. The subsequent rash of selling public organisations to the private sector was based on the economic mythology criticised in this column.

Before leaving on a recent overseas trip to attend two conferences, I made three short telephone calls to confirm arrangements in Washington and London. On one occasion the person I wanted was unavailable but an assistant advised me when I could ring back. That call cost me 30 cents while the two successful calls cost me 60 cents each. All the calls were made from public telephones, all of which actually charged me the correct amount. So the three telephone calls from Australia to the UK and the US cost me $1.50 in total.

I recalled these successful dealings with Telecom when I was in London. From there a call to Australia cost me twice as much as ringing London from here. An earlier encounter with British Telecom was more frustrating. Because they have a system of time-charging for local calls, I ran out of money without actually speaking to anyone when my call to an airline succeeded only in obtaining recorded messages: 'Your call has been placed in a queue', 'Please don't hang up', 'Our operators are still busy' and so on from the mechanical voice as my money ebbed away.

Even that experience almost counted as a resounding success by comparison with my dealings with Pacific Bell Company in California. I had time to make a phone call because of the sort of traumatic experience with American domestic airlines that convinces me of the case against deregulation – but that's another story. My attempt to dial Australia was intercepted by an operator who informed me that the minimum charge for a three-minute call would be $9.10 and asked me if I had the correct change available. The largest American coin in common use, and the largest the phone would accept, is the 'quarter' – 25 cents. To make my minimal call to Australia would therefore have required at least 37 coins, assuming my trousers could stand the strain of that much metal. I gave up and went away to reflect on the palpable benefits of private ownership and deregulation.

We are encouraged by economic zealots to believe that private corporations are inevitably more efficient than public-sector organisations, so the sale of publicly owned enterprises will produce better service for the paying customer as well as easing the burden on the long-suffering taxpayer. If these durable beliefs have any substance, why do the freshly privatised British Telecom and the anciently private Pacific Bell Company provide a service that seems so inferior to that from our own Telecom – publicly owned, extensively regulated and restrained by government from steps it regards as economically rational, such as time charges for local calls? The belief in the inevitably greater efficiency of private companies is one of the hoary myths of our time. It is based on a simplistic view that private corporations are constantly engaged in ruthless competition and so must be efficient or go out of business, while public organisations know they can always pass on to customers the costs of their inefficiency.

The first flaw in this argument is that it confuses control with competition. Competition may produce efficiency and its absence will often allow inefficiency, but that has nothing to do with the question of government or private ownership. There is no obvious reason for a private monopoly to be any more efficient than a public monopoly. Certainly the privatisation of British Telecom has not made it any more generous to the customer or provoked better service. It is clear that time charges for local calls would almost certainly have been introduced in Australia by now if our Telecom had been a private corporation rather than a public body.

The second flaw in the argument is its failure to recognise that each industry has its own characteristics, which are probably more important than the issue of ownership. This is evident from comparisons of different companies in the same industry; would a typical overseas visitor, for example, detect that Australian Airlines is publicly owned and Ansett private? Are there any fundamental differences between the Commonwealth Bank and the private banks? Are state government insurance offices systematically less efficient than private or mutually owned counterparts? These questions would normally only be answered in the affirmative by the most dedicated members of such extremist bodies as the HR Nicholls Society or the Institute of Public Affairs.

The final flaw in the argument is that it implicitly assumes that economic efficiency is the only yardstick by which organisations will be judged. That is fair for some kinds of business: for example, if you always buy the same brand of soap powder, price will be the main factor determining where you buy it. When you patronise an airline or buy clothes or use the facilities of a bank or buy a newspaper, you are making a more complicated decision in which you weigh up the quality of the goods or services against the price.

A further complication is that some organisations, such as Telecom,

Australia Post and the airlines, are seen as having an obligation to the community to provide a level of service to areas that cannot provide the turnover needed for profitable operation. We would not want the postal service or the phone system to go only to locations which generate a profit, any more than we would want to live in a society in which murder was deterred solely by economic incentives.

The clear conclusion is that decisions to sell public sector organisations to private enterprise should not be justified simply by appealing to mindless economic dogma. For each proposal, the case needs to be examined on its merits with due consideration of the economic and social benefits, as well as the corresponding losses. International experience suggests that the supporters of public ownership of Telecom have little to fear from such an examination.

REALISM, FOR GOOD MEASURE

The Weekend Australian, 24–25 August 1991

Before the age of the car, travelling to work was a social event. Most urban commuters lived within walking distance of a railway station or tram stop. We walked along the street in the morning, perhaps meeting our neighbours and talking to them about the weather, sport or other events of the day. If the conversation didn't continue on the train or tram, you could always indulge in the interesting exercise of watching your fellow passengers. Because I usually cycle to Griffith University, I miss out on the social interactions of travel. There is not much chance of conversing even with your fellow cyclists, let alone the motorists who flash past with their windows wound up.

Attending a recent workshop at the Gold Coast reminded me of what cyclists and motorists miss. The sensible way to get there was by inter-city bus. Leaving home about 6.35 on a crisp morning, I found myself sharing the footpath with earnest joggers and power walkers. At the bus stop, I discussed the weather with a couple from Mackay. The bus trip was an opportunity to think over the presentation expected of me at the Gold Coast, but I was doing other things as well. Inspiring music was being injected into my ears by my Walkperson, part of the preparation for my bit-part as about 0.7 per cent of a choir which will sing to the discerning public of Brisbane next week. I observed the passing vista through an enormous window: the used car yards, the surviving remnants of bushland and the tasteless new housing estates, lunar landscapes devoid of trees and grass, decorated with identikit brick boxes.

I was also watching my fellow passengers. A neatly dressed young woman was engrossed in *Dances with Wolves*; you've seen the film, now read the book. Nearby, a male teenager was exposing his mind to much less uplifting material. He eventually left his soft porn to awaken a young woman who had been dozing, curled into a foetal ball on the seat. She stretched and produced some sort of portable decorator's toolkit, from which she proceeded to apply preparations of different colours to various parts of her face before attacking her eyelashes and hair with miniature agricultural implements. A complete work of fiction was created before my startled eyes before she disembarked, appropriately at Movie World, the latest of the rash of theme parks between Brisbane and the Gold Coast.

The bus finally deposited me about 50 metres from my destination. The trip had given me time to reflect on my talk as well as a variety of other matters, from the appalling architecture of what was once a paradise for surfers to the intricate social customs of young Australians. I had about an hour and a quarter of peaceful contemplation on the bus. Had I driven there by car, I would have experienced instead an hour of concentration and stress. I would also have used more fuel, produced more greenhouse gases and spent much more than the $20.50 I paid for my day-return ticket.

This balance sheet came back to me when I read some of the reactions to the draft reports of the Australian Government's Ecologically Sustainable Development process. I agree with some of the criticism of the draft reports, such as the suggestion from the Business Council of Australia that the final reports need to be more realistic, although I know we would not agree on what constitutes realism. A colleague commented that the draft reports had been sanitised to within an inch of their lives. He felt that the reports were so careful to avoid controversy that they had turned out rather bland. If the final reports are to have the impact needed to point us in the direction of sustainable development, there will have to be explicit, concrete and practical recommendations.

As well as the blandness, the reports contained lingering elements of economic romanticism. The working groups agreed to include as one of the elements of sustainable development the notion of improved material and non-material wellbeing. One problem with this goal is its ambiguity. While I interpret it to mean that we should strive for improved wellbeing, taking account of both material and non-material factors, at least one group took it to mean that both material and non-material wellbeing should be separately improved. As if that wasn't bad enough, that group went on to define improved material wellbeing as growth in the GDP.

The GDP is the sum of all economic activities, including those which certainly do not improve wellbeing, such as waste and injuries. In the specific context of my trip from Brisbane to the Gold Coast, I saved fuel and money by taking a bus rather than driving a car. If I had hired a car and driven it to the workshop, the GDP would have increased. It would have been boosted still further if I had crashed the car into a tree and done several thousand dollars worth of damage. By this measure, the nation would have experienced a still greater lift in material wellbeing if I had been hospitalised as a result of driving, and the benefit would have been even larger if I had been made a quadriplegic by the accident.

The obvious point is that we need to have a more realistic measure of community wellbeing if we are going to have a sensible debate about future development. Measuring wellbeing by the GDP is about as intelligent as deciding the quality of a book by the number of pages it contains, or the competence of your MP by his or her weight. With a more sensible measure of wellbeing, we will encourage people to combine pleasure with fuel efficiency, safety and overall responsibility by using public transport. That would be a more realistic approach than trusting economic growth to produce sustainable patterns of fuel use.

WHY LAISSEZ-FAIRE'S NOT REALLY FAIR

The Weekend Australian, 22–23 December 1990

In this traditional season of good spirits, there does not appear much economic cheer. It may be, as I heard said in Brisbane last week, that this is the Christmas we had to have! Still, the retailers do their best to persuade us all to do the right thing at this festive season. Every shop pours forth its flood of music; bel canto gives way to can belto as syrupy versions of European carols assault our ears. Reindeer, sleighs and plastic trees edged with plastic snow form absurd tableaux in the blazing heat of the Sydney summer. Rotund and red-cheeked Santas perspire freely as relays of tiny tots clamber around their padded forms.

While it is tempting to be cynical about the crass commercial nature of the modern Australian Christmas, an extravaganza of reckless spending and gastronomic indulgence, even the most devout atheist can see the value of a season of goodwill to all. It also does no harm to reflect on those things about which we should be cheerful. Fifteen years of the econo-mystic approach has helped to produce the worst economic state most Australians can remember, but there need not be unremitting gloom. Most of us still live in conditions that would be the envy of people in other countries. At this time of the year our tables generally groan with grog and good food, as a result of which most of us eat and drink more than we should. Many of us still own larger and less efficient cars than we need, as well as driving them more often than is wise if we want to live to a ripe old age.

While our cities are undoubtedly less pleasant and less safe than when you and I were young, there are some signs of hope. There are more pedestrian areas now, public transport is being improved and there are serious attempts to preserve some of the remaining urban bushland. Residents of our cities may be prepared to make economic sacrifices for the sake of the quality of life. When the opinions of Brisbane people were sought as part of a strategic planning exercise by the city council, the residents spoke loud and clear. The overwhelming majority of people enjoy their lifestyle and do not want to give it up for material gain. There are enough signs of hope to allow cheer in the heart of someone as congenitally optimistic as me. So what reservations could intrude to mar the season of goodwill?

One is that we are paying a price for present policies in terms of equity. The gulf between rich and poor is widening for a variety of reasons. There is now a tradition of percentage increases in wages and salaries to compensate for rises in the cost of living. The Consumer Price Index (CPI) measures crudely the percentage increase in the cost of living for the average family. However, wages are not increased by the appropriate percentage of the *average* family income; instead, most get an increase of that percentage of their own *actual* wage. So if the CPI goes up by 3 per cent of the average family income we give lecturers an increase of 3 per cent on their larger salaries, while judges and Cabinet ministers receive 3 per cent increases on their much larger pay packets. In fact, the inequity is greater than that. Recent studies have shown that senior executives have been getting larger percentage increases on their greater salaries. Also, those on very high incomes are often able to arrange their financial affairs to pay less tax than many with much lower earnings. As the extreme example, there are individuals who feature in the list of the 200 richest Australians, yet appear to have taxable incomes so low that they do not even pay the Medicare levy!

A more fundamental problem is the effect on equity of applying the principles of the 'free market', the ideology of most members of the economic priesthood which advises government. Adam Smith is often derided for his belief that we are all guided as if by an invisible hand to act in the common good, but he had a more sophisticated view of economic behaviour two centuries ago than some of the present defenders of 'economic rationalism'. I learned the principle of market power nearly 30 years ago, on a cricket tour to Tasmania. It rained one day and eventually the bored cricketers began a game of poker while we waited for the weather to improve. One of our number, with a heavier wallet than others, won regularly, even when we were sure he was bluffing. He could afford to raise the bidding until it reached a level the rest of us could not afford to lose, so we were obliged to drop out and let him win the pot. The same principle applies to the allocation of resources through a market. Those with more money to spend will often be able to secure what they want by bidding up the price, while those in straitened circumstances will lose out. Leaving the allocation of scarce resources to the market allows the government to avoid making hard decisions. This may be acceptable where the resources in question are really frills, like opera tickets or city-centre parking. Where the resources allocated through the market are quality education, with its implications for life-chances, or life-saving medical treatment, it is much more difficult to justify the laissez-faire approach. It may be laissez, but it certainly isn't fair.

The legacy in Britain of a decade of this approach is an economy which is no stronger, with inflation no lower and unemployment no lower, but social divisions that appear much wider. While things appear not to be going well in Australia and the Treasurer has talked about 'the recession we had to have', the economic zealots urge us to have continued faith in the power of the market. There is a clear difference between being cheerful and mindless cheer-mongering. Saying 'we'll all be rooned' is not a rational response to hard times, but neither is a child-like trust in market miracles. The market may well deliver wealth, but it is likely also to make us a less equitable country. Is that really a Christmas present we want?

TAKING A LEAF FROM THE DARK AGES

The Weekend Australian, 3–4 November 1990

There are several things we need to do to become a clever society. Developing the capacity of our libraries to provide an adequate information base is one of the most urgent. The 1990 Science and Technology Statement argued that we need a scientifically literate population to understand the role played by science and technology in the economy. It also said that scientific literacy is needed for informed debate about the ways in which we might use the scientific and technological opportunities before us. So the government sees scientific literacy as being essential if our economic decline is to be halted, as well as being the prerequisite for informed public debate.

The importance of adequate libraries stems directly from the scale and pace of change. We cannot give today's students the skills and the knowledge they will need for their working lives, because much of that knowledge and many of those skills do not yet exist. In a time of rapid change, formal education must be seen as the first step in a process of lifelong learning. Lifelong learning requires continuous access to information. The Hon. Barry Jones, former Commonwealth Minister for Science, is now chairing a new House of Representatives standing committee on long-term issues. As its first task, it is looking at the question of a national information policy. I wish the committee well, although historical experience suggests that producing a good report is only part of the battle. In this same area, the Horton report on public libraries was an excellent document with many important recommendations. Some 14 years and six federal governments later, we have yet to see any of those recommendations implemented.

As one feature of this discussion, we should be developing mechanisms for a public input to the overall information strategy. Some would say we should leave this to market forces, but there is a serious drawback to that approach. An entrepreneur who is paid 100 times as much as the national average effectively has 100 times as much say about which goods are on the market. Even if you accept that with good grace, you might worry about the idea of the great entrepreneur having 100 times as much say as the average person about what information will be made available to whom.

Decisions that appear to be simple technical choices are often really about the equitable allocation of resources and equitable access to information. About five years ago, Professor Graeme Duncan suggested some sound operating principles as a basis for planning. It is an obligation of government, for example, to ensure equitable access to information. Since access to information is increasingly important for social and economic mobility, it is unacceptable for that access to be determined by such arbitrary factors as the care with which our parents chose their postcode. In terms of social equity, we also need to develop ways of improving the access to information of those in groups that have traditionally been disadvantaged: women, recent immigrants, older people, Aboriginal people and Torres Strait Islanders, all those in remote areas and members of relatively poor communities.

James Maddison was quoted in Barry Jones's book *Sleepers, Wake!*

as saying that a popular government without popular information or the means of access to it is the prologue to a farce or a tragedy, or perhaps both. 'Knowledge will forever govern ignorance,' he said, 'so people who mean to be their own governors must arm themselves with the power that knowledge gives.' This is an important point. Information in modern society is equivalent to political power. If people are to be involved in decisions about new technology or industrial initiatives, we have to ensure that they have the information needed for that involvement.

Although we have known for years that we need to take steps to improve access to information, little has been done in concrete terms. Worse, we have recently seen the deployment in this area of some of the economic ideas that are doing so much damage in other areas of modern society: the user-pays principle; the alleged benefits of privatisation; the purported value of small government; and the belief in economies of scale. The application of these ideas by people who know the price of everything and the value of nothing is moving us inexorably toward a new age of barbarism. We are seeing an erosion of the infrastructure that would be necessary to provide for the needs of the traditionally disadvantaged.

Our libraries are our windows on the knowledge of the world. There has been a steady run-down of libraries in the past decade as their funds have failed to keep pace even with inflation, let alone the ballooning costs of publications and the burgeoning demands of the information society. Letting this happen amounts to an acceptance that we will live in a new Dark Age in which we will be increasingly unaware of developments in the rest of the world.

I do not believe we should meekly accept that fate. The libraries of the future need to be better funded to meet broader needs. Libraries have evolved in recent years from repositories of books and journals to include films, videotapes, other audiovisual materials, access to databases and online information services. The demands we will be making on these information centres in the future require that they be properly funded. Governments need to be aware that their actions, or their studied inaction, are profoundly influencing the possibility of providing an adequate information service for the clever society. A proper information policy is an integral part of the process of moving toward that goal. Resources must be provided to enable development and implementation of a national information strategy.

PAUSE TO SEE THE FOREST HOME BEHIND YOUR TIMBER HOUSE

The Weekend Australian, 18–19 May 1990

How would you feel if the agents of a foreign country were destroying your local shopping centre? Would you see their looting and pillaging as a small price to pay for progress? Or would you angrily demand to know why your government was allowing such an outrage? Would your anger subside at all if you were told that the local destruction was helping to generate export income and therefore improve trade figures? I don't think so. How upset would you be if you were told that local politicians were profiting from the operation?

I don't imagine many readers of this newspaper would be content to stand by and watch their local supermarket and other shops being ravaged, whatever the alleged national benefits. The shopping centre analogy was suggested to me by Dr Bob Brown in a speech he made as part of the campaign to stop the import of tropical rainforest timbers. To the people who live in the forests of South-East Asia, those complex natural systems are their source of food, clothing, building materials and medicine.

The forest is mainly timber. We effectively offer such countries as Malaysia and Indonesia financial inducements to log their forests. We are actively collaborating with them to destroy the forests, which are not only natural systems of great value but also the resource base of the native people. We are bulldozing, logging and burning their shopping centres.

The loss of forests is a significant contribution to the increasing carbon dioxide concentration which is causing climate change. Some Western authorities have even gone so far as to suggest that the main cause of global warming is the clearing of tropical forests. While this claim is a dishonest attempt to justify continued profligate use of fossil fuels, there is no doubt that the loss of forests is a serious problem. Worldwide, we are reckoned to be losing forest at the rate of about a football field a second. While you have been reading this column, an area about half a kilometre square has been cleared. We should not be supporting such wholesale destruction. That is why the issue of Australian imports of rainforest timber is so important.

A pamphlet printed by the Rainforest Information Centre at Lismore has suggested that this level of logging cannot be continued for much longer, as the resource is being rapidly depleted. It says that, by 1995, much of the forest area in our region will have been logged, an area of rainforest four times the size of the United Kingdom will have been lost, and depletion will have set in train the collapse of the tropical timber industry.

The World Resources Institute has estimated that the majority of the nations which now export tropical timber will be importing wood by the year 2000. By then, on present trends, almost all of the unprotected forests of South-East Asia will have been logged, except those in Papua New Guinea. The lifestyles of most of the indigenous peoples living in those forests will have been irreparably damaged in the process. It has also been claimed that almost half a million species of plants and animals will have been made extinct. Though I can't see how this figure can be known with any precision, there is no doubt that there will be massive loss of species diversity if clearing of the forests continues.

There are some specific decorative uses for beautiful timbers from rainforests, but much is squandered on throwaway items. I have been assured that there are adequate substitutes for the rainforest timbers we now import. For many of the applications, plantation pine is a reasonable replacement and one we can use with a clear conscience. So why do we continue to import rainforest timbers?

One reason is the manic commitment in Canberra to free trade. Many government advisers are very reluctant to recommend anything which could be considered as a restraint of trade, lest their devout belief in 18th-century economic orthodoxy be questioned. A second possible reason might be a feeling of national guilt. Having destroyed much of our own rainforest, we can hardly be self-righteous about other countries clearing theirs. That is true. We certainly have no right to make moral judgments about those who are sufficiently desperate to be prepared to sell their heritage. But equally, even if we now see our forebears as having been irresponsible in the way they treated rainforests, that hardly obliges us to encourage others to behave in a similar way.

Another explanation could arise from a genuine concern to help the economic development of countries to our north by agreeing to buy whatever they have to sell. While that commitment does not appear to be applied consistently, it can't be denied that we do have some obligation to assist the development process. The key issue is whether we are doing countries even poorer than ours a favour by encouraging a pattern of development which cannot be sustained.

I have recently come under heavy fire from Australian forest industry officials, most of whom believe that their local activities are sustainable. That claim can't be made seriously about the logging operations in South-East Asia. If we are really concerned to help our neighbours to the north we should be putting together packages of aid and trade which encourage sustainable patterns of economic development. In that way we could help to promote responsible development strategies in the countries of South-East Asia. Encouraging the destruction of their natural shopping centres is not the action of a good neighbour.

IT'S TIME TO SEE THE FORESTS, NOT MERELY THE TIMBER

The Weekend Australian, 7–8 September 1990

This week has seen a good old-fashioned punch-up between the forest industry and the Minister for the Environment, Ros Kelly. It temporarily distracted attention from our brave ministers standing firm against appeasement in the Gulf, an image which would have been more impressive if they weren't simultaneously caving in to the economic zealots at home by trying to flog off any profitable public asset that isn't nailed down.

The fur started to fly when Ms Kelly said that the forest industry's aim was unrestrained development, accusing them of wanting to amend the Australian Heritage Commission Act to gain access to more of the nation's forests. The industry's reply was distinctly unconvincing. The executive director of the National Association of Forest Industries, Mr Bain, conceded that the industry did want to amend the Heritage Act 'as a way of achieving resource security to encourage investment'. That sounds remarkably like gaining access to more forests to me.

Mr Bain defended the industry against the charge that they favour unrestrained logging. He said that they had endorsed the prime minister's sustainable development discussion paper. That paper gives every appearance of having been drafted by an interdepartmental committee dominated by economists. It starts from a commitment to continue economic growth and makes only token concessions to sustainability. As it has been condemned by conservation groups and endorsed by various industry groups, supporting it could hardly be seen as proof positive of a commitment to ecological sustainability.

Forestry is one of the most important issues in resource politics. A recent Canberra conference on policies for moving toward global sustainability devoted a whole session to forestry, with speakers from the industry, conservation groups and academia. The proceedings of the conference will be published later this month by the Australian National University. Neil Byron and Robert Davies pointed out that the key problem is deciding what is to be sustained. Often the industry seems to equate forestry with the production of timber. It has been said that an accountant is someone who can't see the wood for the trees, while a forestry accountant can't see the trees for their wood.

In those terms, forestry would be considered by some people to be sustainable if the volume of marketable timber stayed constant, even if the old eucalypt forests were steadily being replaced by hectares of pine trees. It is entirely possible to have a pattern of timber production which is economically desirable but not ecologically sustainable.

I have been very impressed by many of the foresters I have met. They clearly understand our forests serve a range of functions. As well as being a source of timber, they also protect the soil, preserve a habitat for flora and fauna, absorb carbon dioxide from the air, influence climate and serve as a community resource for experiences ranging from a short stroll to a serious wilderness trek. Our forests play a crucial role in maintaining biodiversity.

It is important to recognise this variety of roles played by our forests. It is no more reasonable to see them solely in terms of timber production than it is to see Fraser Island solely as a source of sand minerals. The wilderness values of Fraser Island are now seen, by all but a few determined troglodytes, as sufficiently important to forgo the possible economic benefits of mining. In similar terms, there are some areas of forest which most people would now want to see protected from the axe and the chainsaw.

While those forests produce tourist income, the main economic benefits from forestry still lie in timber production. The pattern of our trade in forest products is quite startling. In 1987–88 we imported a total value of $1.99 billion and exported $0.48 billion. In other words, our imports were worth four times as much as our exports, giving a trade imbalance of $1.5 billion. By value, over 70 per cent of our imports are in the form of pulp and paper products. Perhaps more surprising was the fact that the same proportion of our exports was in the form of woodchips.

This is an important observation because, as Byron and Davies pointed out, it is sometimes argued that the local industry cannot supply our paper needs because conservation groups are denying use of our forests. The industry statement earlier this week was a clear call for access to more forests to give 'resource security'. But is this really the key problem? If access to our forests was being denied by green activists, would we be exporting $350 million worth of woodchips annually? The clear inference is that our forest industries lack processing capacity rather than access to wood.

We appear to be importing value-added products from Canada, New Zealand, Scandinavia and even southern America while exporting cheap woodchips, often from old-growth forests. It has been suggested we must be undervaluing our forests if it makes economic sense to sell them as woodchips. Bob Davies told the Canberra conference that the price of wood at the mill door would have represented only about 7 per cent of the operating cost of the proposed Wesley Vale mill. As further evidence to support the view that our timber is undervalued, some state forestries seem to be subsidised operations. A recent study of the Forestry Commission of Tasmania reached the conclusion that the loss on its operations over the past 50 years has been over

$650 million. Recognition of the scale of traditional subsidies has led to recent enormous increases in some royalty charges.

In terms of slowing down global warming, storing carbon for decades by using wood as a building material or for furniture makes much more sense than shredding it or burning it. We should encourage responsible use of wood. More generally, we should encourage our foresters to continue managing those resources so that the forests themselves are genuinely sustainable. New institutions like the Tasmanian Forestry Council, bringing together the different interest groups to negotiate an industry strategy, may be the structural key to developing a plan for management of these important resources.

A VOTE FOR ACCOUNTABILITY

The Weekend Australian, 1–2 June 1990

I am worried about the final report of the Fitzgerald Inquiry on Fraser Island. This is not to say that the report has obvious deficiencies. On the contrary, Mr Fitzgerald did a commendably thorough job.

As an example, the inquiry took a much more considered approach than we usually see to the vexed issue of logging. The report drew a clear distinction between the logging of different types of trees. While it found that forests containing blackbutt trees are disturbed by logging, it saw no evidence that the natural ecosystem is being threatened by logging at the current rate. Hence, it recommended that harvesting of blackbutt from previously logged areas could continue as part of an overall forest management strategy.

The inquiry was unable to reach a similar conclusion about brush box and satinay forests, expressing significant doubt about the ecological consequences of logging at the 1990–91 rate. It adopted the cautious approach of saying that the activity should be ceased, since its sustainability cannot be established.

The report recognised that a ban on the logging of these species will have serious economic effects in the Maryborough area, where each of the two sawmills gets more than half of its timber from Fraser Island. There is also speculation that the continued logging of blackbutt alone won't be an economic proposition. It is unclear why the data are not available to answer that economic question.

Mr Fitzgerald recommended that there should be compensation to those affected by the proposed ban on logging brush box and satinay forests. That seems both sensible and just to me. Clearly the public purse should compensate those affected when changing public values leads to a change in resource policy. We do need to ensure that the compensation payments reach the workers whose livelihoods will be affected.

The report was equally cautious in its attitude to the revival of attempts to secure permission for sand mining. This proposal came as a bit of a shock. Most people thought that the issue had been resolved by the Fraser Government more than a decade ago. Some people never give up! On the basis of its inquiries, the report could not conclude that the ecosystem can be restored after sand mining. It took the cautious line that mining should not be allowed, suggesting that the issue should be resolved once and for all by the termination of existing mining leases on the island.

The report also looked at the ecological and economic impacts of tourism. This is a key issue, since feral four-wheel-drive vehicles do considerable damage to Fraser Island. It could be argued that they have done more damage to the environment than logging operations. I was surprised to learn that the fees collected on behalf of the Queensland Recreation Areas Management Board in 1989–90 from more than 200,000 visitors only amounted to about $6 each!

We should have a proper management scheme for an asset such as Fraser Island. The report concluded that it meets all of the prescribed conditions for World Heritage Listing. The island is genuinely unique, in the literal sense of that much-abused word, and should be protected from any form of exploitation that would seriously threaten its natural values. Those who enjoy Fraser Island should contribute to its protection.

With so much that I see as good in the report, you might be wondering what worrying signs I found in it. I think that the process it has used sets a dangerous precedent. Essentially, the inquiry conducted a range of studies and sought evidence from all interested parties. It did not use any sort of adversarial process to resolve differences between competing viewpoints, such as the conflict about the long-term sustainability of logging. Instead Mr Fitzgerald, as one wise individual, weighed the evidence and reached his conclusions. These were not simply evaluations of the scientific evidence about sustainability, but involved a balance of various competing interests, such as economic exploitation and conservation of natural values.

The difficulty I have with this process is that I do not think it is reasonable to ask one individual, or even a group of non-elected people, to take such decisions on behalf of the community. The balance between economic and ecological demands is a political one of assessing community values. I believe it should be taken by the political process.

Contrast the Fitzgerald approach with that taken to the issue of Coronation Hill by the Resources Assessment Commission (RAC). Their report has been attacked on two grounds. The method used to quantify the natural assets of the Kakadu region came in for some understandable criticism. Asking people who do not have to put their hands in their pockets how much they would be prepared to pay to protect Kakadu is of limited value. A much more convincing economic argument against the mining of Coronation Hill was the derisory potential return per head of population. If the return would have been, as estimated, about $5 for each Australian, that hardly seems to justify mining in a wilderness area of considerable natural value.

The more fundamental argument against the RAC was that they didn't make a concrete decision or a clear recommendation. I believe that the RAC approach was totally correct. They spelled out the competing interests: economic, ecological and anthropological. The balance between these competing interests is a political one and should be taken openly by our politicians rather than by faceless persons, however wise and incorruptible.

If you don't like the decisions a government makes, you can legitimately campaign to transform them into the Opposition at the next election. If one wise person's view of a complex land-use decision is not the one you would have taken, to whom do you go to appeal against the decision?

THE 2020 SUMMIT

A report prepared for the Australian Conservation Foundation, May 2008

The 2020 Summit was an inspiring event, with enormous energy and strong commitment to the possibility of shaping a better future. Overall, I thought there were many more positives than negatives and we could be very pleased with the level of support at the 2020 summit for the sort of future that we want, not just in the specific section that I was involved in. In Kevin Rudd's closing speech, he said:

> There is no single policy which can respond to every aspect of climate change. Rather, the Summit discussion suggested almost every aspect of policy choice will be affected by the drive toward sustainability. Climate change is the overarching issue this generation and those to follow must address . . .

I was in the population, climate change, sustainability, water and the future of cities stream. It concluded:

> Australia faces an unprecedented challenge from climate change. We risk losing our natural heritage, our rivers, landscapes and biodiversity. We have a brief opportunity to act now to safeguard and shape our future prosperity.

Our section went on to set out some broad ambitions:

> Our aspiration is that by 2020 Australia is the world's leading green and sustainable economy, that we will set time-bound targets and be on track to dramatically reduce our ecological footprint . . .
>
> By 2020 Australia will be making a major contribution to a comprehensive global response to climate change, including working with our partners on clean energy. Australia will have dramatically reduced our emissions . . .
>
> Environmental considerations will be fully integrated into economic decision making in Australia at the household, business and government levels . . .
>
> A robust emissions trading system and a suite of complementary measures will be driving a low-carbon revolution with Government taking the lead, working in partnership with business and the community. Climate and sustainability policy will also incorporate the needs of disadvantaged and low-income Australians.
>
> A new dialogue will have been established with our Indigenous peoples on our response to climate change, water and sustainability challenges.
>
> Australia's globally outstanding ecosystems and species are managed to reduce threats and build resilience to promote adaptation to climate change.
>
> By 2020 the health of Australia's ecological systems will be improved . . .

We advocated an integrated, whole-of-government approach underpinned by clear targets and measurement with independent reporting. We also identified strong national leadership and international engagement as high priorities. We suggested a National Sustainability, Population and Climate Change Agenda with robust institutions to support it. We hoped that Australia would develop a whole-of-government approach to climate change and sustainability policy, encompassing government expenditure, taxation, regulation and investment, including an audit function to report on government's performance against these climate change and sustainability objectives. The section also advocated fundamental economic reform, such as a set of national environmental accounts, including carbon and water accounts, to inform government, business and community decision making.

The most disappointing aspect of the Summit was the response when I proposed a formal statement that no new coal-fired power stations without carbon capture and storage should be built. I prefaced this proposal with an explanation that there were different views around the room about whether carbon capture and storage was technically feasible, economically practical, ecologically viable or socially acceptable, so there would be different views about whether coal-fired power stations with carbon capture and storage might be acceptable, but we could surely all agree that coal-fired power without carbon capture and storage is now unacceptable! Remarkably, that was not the case, and the troglodytes in the coal industry and the economics profession opposed this proposition, necessitating Senator Wong reporting that 'a substantial number of the group felt strongly' that there should be no more old-fashioned coal-fired power stations but 'there was no consensus'. While there was a strong consensus on the need for a rapid transition to clean energy supply technologies, there were some at the Summit who believed that this label included 'clean coal' or even nuclear power! While the sub-group that I was in agreed a consensus position that we must 'dramatically reduce our ecological footprint while striving to maintain or improve quality of life', the final statement after including the views of the other two sub-groups said 'dramatically decrease our ecological footprint while continuing to grow our economy and improve our quality of life' – so the old approach that quality of life is improved by economic growth is perpetuated in the statement. I concluded that there remains in Canberra a pervading neo-liberal ethos that sees price signals and market forces as the main driver of change, with regulation and government action as supporting strategies.

While there was grudging acceptance that we should have a population policy, there were clearly very deep divisions about what that policy should be. I argued that we should have a goal of stabilising the population at a level that could be sustainably supported, but others opposed even that sort of general proposition that allowed room for debate about per capita consumption levels. Many of the people in that group clearly still hope that rapid technological change could somehow compensate for the increasing demands of a growing population.

THE NEW CONSENSUS

The Age, January 2009

The Washington consensus is dead. Is a Zurich consensus emerging? In 2008 similar conclusions emerged from three major international conferences convened by totally different organisations, all based in Switzerland, and a report from a major international agency.

The Geneva-based World Economic Forum (WEF) would normally be considered a conservative body. The reality of the financial crisis concentrated the minds of participants in their Summit on the Global Agenda in Dubai early in November. They concluded that the world economy can't go on as before. The word that kept recurring was 'reboot' – start the system afresh. Many saw the fundamental problem as the widening gap between financial trading and material reality. With the monetary value of financial transactions now about a hundred times the value of traded goods, the whole house of cards was bound to collapse sooner or later.

A broader critique emerged during the Summit. Successive reports on the world's social and environmental conditions have been ringing alarm bells for more than a decade. The warnings were ignored by leaders who simply concentrated on the economy. In Dubai Professor Karl Schwab, head of the WEF, said that the financial crisis could not be considered in isolation from the other great challenges of climate change, energy security, water and food. In other words, we need an integrated policy approach that recognises biophysical and social reality as well as economic opportunities. We can no longer assume that economic growth will solve the problems, he said.

The Earth Dialogues in Brazil started from the other end but reached a similar conclusion. The conference was organised by Zurich-based Green Cross International in association with the government of Minas Gerais, a Brazilian province with a strong mining emphasis. The conference concluded that water is a social need and its protection an environmental imperative. It also recognised that providing clean water in poor countries is compromised by the financial crisis. The Millennium Goals of clean water and adequate nutrition won't be achieved unless we deal with the economic development needs of the poorest nations.

Until recently the International Energy Agency has been cheerfully promoting business as usual, talking about world energy use doubling by 2030. But in its highly anticipated 2008 World Energy Outlook, also released in November, it said:

> The world's energy system is at a crossroads. Current global trends in energy supply and consumption are patently unsustainable – environmentally, economically, socially. But that can – and must – be altered; there's still time to change the road we're on. What is needed is nothing short of an energy revolution.

We have known about the interlinked problems for more than 30 years because of the fourth group. The Club of Rome, also based in Zurich, became famous in 1972 when they released *Limits to Growth*. Now two different studies, one by Australia's science body CSIRO and the other by US economists, have compared the 1972 projections with 30 years of data. Both found we are following exactly the path projected by the Club of Rome. We are right on track for collapse.

The Club of Rome also convened a November conference on managing the interconnected challenges of climate change, energy security, ecosystems and water. It concluded that the problems interlock and demand an integrated approach. The most urgent challenge is climate change. Almost every week there is more scientific evidence showing that change is accelerating and we risk catastrophic disruption. We still hear some irresponsible calls for a weak response, claims that we can't afford to tackle climate change with the financial system in turmoil. This sort of talk is not just out of touch with the biophysical reality, it is also completely at odds with the emerging global view of economic development. The Zurich consensus is that an integrated approach to economic, social and environmental problems provides the only chance of preserving civilisation.

In ecological terms, we are on the Titanic and heading for the iceberg. It is irresponsible to steer straight ahead or throw more coal in the boilers. We need to plot a different course. The situation is too serious to be pessimistic or scared into denial. When the World Economic Forum, the International Energy Agency, the Club of Rome and Green Cross International agree, it is time for our leaders to sit up and take notice.

CULTURE AND HEALTH

INTRODUCTION: AUSTRALIAN CULTURE AND THE PROSPECTS FOR A HEALTHY FUTURE

Anyone who has travelled overseas knows that we have a unique Australian culture. For that matter, you only have to watch film or television from other countries to appreciate the differences. Culture has great capacity for endurance. In the 1960s, when Mao Zedong was leader of China, he urged his rabid young followers, the Red Guards, to tear down all the trappings and institutions of the old China. Great upheaval occurred: teachers and intellectuals were publicly humiliated; ancient treasures were destroyed. At the height of the hysteria, the Red Guards greeted Mao with rapturous shouts hoping that he would live 10,000 years – exactly the way emperors had been hailed in the feudal era!

On the other hand, there is no doubt that cultures can change and do change. The Australia of today is very different from the Australia of my childhood. We are, in many ways, a healthier society. We live longer, partly because the infectious diseases that killed many young people have largely been eliminated, partly because modern medicine is able to save people who would have died 50 years ago. But we have a whole new set of health hazards, some a function of our living longer and others a result of lifestyle changes. The writings I have collected together in this section are intended to illustrate the complexity of our culture, its enduring features and the capacity for change, as well as the factors that influence our health and the prospects for a healthier society. As health is not simply an absence of illness but also related to general wellbeing, it is affected by social influences as well as physical factors. So our culture and lifestyle have a large influence on the healthiness of our society.

AUSTRALIAN CULTURE AND THE PROBLEM OF ACHIEVING SIGNIFICANT CHANGE

Edited extract from A Big Fix, *Black Inc. 2005, 2009*

One of the endearing features of Australia is our unique culture: our robust sense of humour, our irreverence for authority, our poetry, drama and fiction, our music and our art. Beginning as a reaction against our British colonial authorities, our culture has gradually broadened as successive waves of migration have enriched our society. In a range of areas, our cultural traditions have disproportionately made an impact on the world; our actors, singers, writers and filmmakers are much more visible on the world stage than you would expect of a country of 22 million people. When I was young, most of our writing and film-making was aggressively Australian, telling our stories in our language. In some ways, this represented an over-romantic representation of 'the bush' as if it were Australia epitomised; even in the 1950s, more of us lived in the cities and regional towns than in rural areas.

The increasing control of our cultural institutions by overseas interests has begun to erode Australian culture. Our cinemas show predominantly US films, our commercial television is also dominated by US material and a US media baron owns most of our daily newspapers. The extreme example of Rupert Murdoch demonstrates most vividly the insensitivity of our politicians to this issue. Under US law, Murdoch was unable to buy a major media outlet like the *New York Post* because he was not a US citizen. So he renounced his Australian citizenship and became an American, secure in the knowledge that our law-makers would have no problem at all with our print media being dominated by a US citizen. As an American, he controls most of our newspapers, meaning that no government dare risk his ire by curbing his disproportionate power. His media outlets pursue a simplistic economic agenda, promoting 'free trade' and unlimited growth as the essential elements of the policy program.

Previous protection for Australian actors and other media practitioners will now be systematically dismantled as part of the so-called 'free trade' agreement with the US – really an open door for US companies in return for a few minor concessions to Australian farmers. The recent book by Linda Weiss, Elizabeth Thurbon and John Mathews, *How to Kill a Country*, documents this in depressing detail. Increasingly, TV advertisements are not even dubbed with Australian voices but simply broadcast as if we were an American colony. The suburban milk bar has been gradually replaced by US junk-food chains, using their huge marketing budgets to persuade gullible consumers to pay more for an inferior product. The trade deal with the US means that the situation will get steadily worse, because it limits protection of local content to arrangements that were in place before the agreement was signed. So the limited provisions for some Australian content in existing media outlets will continue, but there will be no provisions at all for Australian content in new media still to be developed. Our capacity to tell our own stories in our own voices will be steadily eroded, leading to a tendency for young Australians to model themselves on the US characters in TV or films. Young

Australians are already more familiar with US cartoon characters like Homer Simpson than with our local real identities. We are now seeing, as an extreme example of the influence of US television, young footballers aping the American habit of clasping a hand to their chest when the pre-match national anthem is played.

Probably the most insidious result of this media control is the increasing emphasis on consumption. Television is a very effective marketing tool, increasingly used to persuade us to buy things we don't need or even want. Even when there is an effective documentary on TV about the state of the world and the problems caused by unsustainable consumption levels, its message is swamped by the surrounding commercials urging us to consume. While globalisation has some positive aspects, one of its disadvantages is the gradual erosion of our distinctive culture by the insidious coca-colonisation of the US media. A second consequence of that is the increasing level of community violence. A typical child has probably seen hundreds of murders on TV and is quite accustomed to seeing police and security officers wielding guns.

Those examples show that cultural change can be achieved, if enough resources are devoted to the task. The McDonaldisation of the Australian suburbs, converting us to patrons of a series of US junk-food chains, was driven by a huge marketing budget. So how could we change to a sustainable future? It is no small task to achieve significant change at any level – the personal, the household, the small community or the nation. Most of the time, most people are reasonably content with the way things are and reluctant to embrace change, especially at a fundamental level. I believe that there are four steps to major change: discontent, a new vision, viable pathways and commitment. This analysis is developed further in a later essay, 'Changing public attitudes to long-term issues' (page 182).

Unless there is discontent, there is no motivation to change. Being discontented is not enough. People are reluctant to change until they can imagine a better alternative. Without a vision, change might actually make things worse rather than better. Again, even a coherent vision is not enough if we can't see a way of getting there from where we are. So the third component of change is developing and describing feasible pathways to get from where we are to where we want to be. The fourth component is commitment. If I can spell out a coherent program of diet and exercise that will recover my physical fitness, it will still take commitment to stick to the regime.

The crucial first step will be committing to the principle of a sustainable future. It should not be necessary to say this, because way back in 1992 COAG adopted the National Strategy for Ecologically Sustainable Development, but there is little sign of these ideals today. We should be flexible about how we get there, but the goal must be unambiguous.

There are several obstacles to the sorts of changes needed to bring about a sustainable future. An obvious one is the dominance of short-term thinking. Apart from the effects of global climate change, none of the serious consequences of unsustainable living are likely to happen while this generation of politicians is in office. As the historian Paul Kennedy observed, people who succeed in democratic political systems are usually those who avoid antagonising powerful interest groups. This means they will not make difficult decisions now in the interests of future generations, as long as they can argue that the experts are divided and more research is needed. With complex questions like climate change, peak oil and world population, there will always be some differences of opinion between experts and an obvious need for more research, so there will always be an excuse for what Sir Humphrey Appleby called 'a strategy of masterly inaction'. Politicians tend to stick to a course of action once it has been announced, because of the notion that changing their mind is a sign of weakness. As most politicians are now committed to a road that is obviously not sustainable, it is hard to persuade them to change and admit that their old ways were wrong. It does happen, however. The significant environmental advances of the last 30 years have been linked with some fundamental reversals of attitude: in regards to big dams in Tasmania; sand mining; logging of old growth forests in most states; the need for rivers like the Snowy to have enough water to maintain their ecological functions; unlimited fishing and so on. As Keynes famously said, we should change our mind when new evidence makes it sensible to do so, rather than pig-headedly sticking to our old conclusions and habits.

A second obstacle is incomplete knowledge; we have probably identified only 10–15 per cent of the species that inhabit Australia, so we can't possibly have the understanding to ensure that we are not making more mistakes. We need serious investment in the new field of 'sustainability science', analysing at the level of complete systems and taking into account the uncertainty and subjectivity of our knowledge.

The third obstacle is technical hubris. Since science and technology have miraculously expanded production from natural systems to cope with the dramatic population increases of the last 100 years, some people tend to believe that there will always be a technical solution to any problem. Those who believe this see no need to reduce wasteful energy use because they are counting on a technical fix, like developing the capacity to capture carbon dioxide and store it in geological formations. Likewise, the water shortages in Western Australia were seen not as an opportunity to initiate more careful use of water, such as growing native plants rather than water-thirsty exotics, but for new supply options, such as desalination of sea water or even bringing water from thousands of kilometres away.

The fourth obstacle is econo-mysticism. Some economists and their followers in politics and the commercial media have a touching faith that suitable price signals can solve any problem. There are, of course, situations in which pricing can influence behaviour, and it is foolish to ignore these possibilities. Equally, it would be silly to believe that the sudden demand for parachutes if a plane malfunctioned would produce a supply before the inevitable crash! In other words, we have to be aware of the timescale needed to respond to a need for change. It is just silly to believe, as one prominent economist actually said recently, 'If the price is high enough, even roosters will lay eggs.'

A variant of econo-mysticism was labelled 'cheer-mongering' in the 1970s. The approach of the cheer-monger is simple. You take a complex issue, such as global warming or the peak of world oil production, and reflect on its complexity. You then announce that there isn't a strong case for the pessimistic view that we should reduce fuel use to slow down climate change, or that we should prepare for the approaching peak of world oil production. Unlike those who preach gloom and doom, the cheer-monger says, 'We have faith in the human spirit and our capacity to solve problems. So be cheerful and consume!' The rhetorical trick is to cast this approach not as naive optimism but as the considered outcome of deep reflection. To the cheer-monger, facts are an inconvenient complication that must be swept aside. So as I wrote, business interests were still calling for a massive increase in migration to boost the economy, without any recognition that the demands of the present population are already irreparably damaging the natural systems of Australia. Journalists in the commercial press were urging the government not to pander to uninformed prejudice by taking any serious measures to reduce greenhouse gas emissions. One right-wing pressure group even claimed in 2005 that the Murray is in perfect health, despite the vast body of evidence to the contrary, on the basis of one indicator – salinity – that has improved in recent years.

This brings me to the final problem: the role of the mass media. The Menzies Government's decision in the 1950s to allow television to be a largely commercial activity has allowed it to become a powerful mass marketing tool. The commercial media is very reluctant to antagonise their advertisers, so they tend to take a conservative pro-business approach. The economics of the print media is also dominated by advertising revenue rather than the price paid for newspapers or magazines, so they also subtly promote growing consumption. At a population conference in Canberra in 2004, a senior journalist argued that population growth was good because it boosts the economy. A questioner pointed out that OECD statistics showed a significant negative relationship between population growth and economic output per person – in other words, the countries that performed best on the indicator of increasing wealth per person were those with slow population growth or declining population. The journalist responded by stating that he hadn't seen the OECD data but the figures didn't sound right to him and he was sure the people he talked to would not agree with them! Hence, when the facts are in conflict with one's prejudices, the facts are discarded rather than the prejudices. With such an attitude, there is little chance of more considered views being aired in our media.

POPULATION AND PROSPERITY

In 1996, this was printed in The Courier-Mail, *summarising the argument in* Understanding Australia's Population Debate, *a booklet published by the Bureau of Immigration and Population Research.*

Continuing population growth is not necessary or desirable. We should aim to stabilise the Australian population at a level that allows sustainable living standards and protects our unique environment. This goal would not mean shutting the door on migrants. We could still be generous to refugees, keep a modest family reunion scheme and provide for limited numbers of skilled migrants. But we need to get rid of some myths that have prevented a serious population debate.

It is not true that our population is growing mainly because of migrants coming here. The annual migrant intake has varied over the last 30 years from 30,000 to 150,000 with an average of about 75,000. The annual 'natural increase' – the difference between births and deaths – is consistently about 120,000. So the main reason our population grows about a million every five years is the birthrate. It is now falling slowly because we are an ageing society. But even if we stopped all migration tomorrow, our population would still keep growing for another 30 years.

There is no truth at all in Pauline Hanson's claim that migrants are mainly from Asia. About 4 million Australians were born overseas. By far the largest group, about 1.2 million, is from the UK and Ireland. The next largest groups – about 250,000 each – are from New Zealand and Italy. Many more people have come here from Greece or the former Yugoslavia than from Vietnam, the source of the largest group of Asian migrants. Many more have come from Germany or the Netherlands than China, the source of the second largest group of Asian migrants. The One Nation party is not really opposed to migration; some of its candidates for the Queensland election were migrants. They only object to Asian migrants – and then wonder why they are seen as racists!

It is also a myth that migrants take our jobs and increase unemployment. Migrants increase the demand for housing and other goods. These needs create about as many new jobs as the migrants fill. In fact, most economists favour high levels of migration as an economic stimulus.

The one serious reason we should be concerned about population growth is the pressure it puts on our environment. The 1996 national State of the Environment report showed we have a range of serious problems such as soil loss, degradation of rivers, land contamination and threatened species. These problems result from the combined effects of population growth and distribution, our lifestyles and the technologies we use. In south-east Queensland, population growth is directly causing loss of natural vegetation and worsening urban air quality. Twice as many people means twice as many cars polluting the air, twice as much bush cleared for housing and twice as much sewage going into the ocean. The ABS recently predicted that the Queensland population will almost double by 2050 if present trends continue. That would destroy the quality of life we have now.

BRINGING ART AND SCIENCE TOGETHER

Prepared for Adelaide Festival of Arts 2002 to accompany the exhibition conVerge

Achieving a creative synthesis of the arts and sciences is vital for the future of human civilisation. We face huge challenges of trying to meet the growing material expectations of an increasing population, from the planet's finite resources and limited capacity to handle our wastes. Most of the complex issues we confront transcend the traditional academic disciplines and cross the boundary between British scientist CP Snow's two cultures. Snow lamented the fact that science and the arts were two distinct worlds and that while most scientists had some knowledge of and respect for the arts, most arts graduates had neither knowledge of nor respect for the natural sciences. Solving the problems before us will require a melding of scientific knowledge, technological capacity, craft skills and artistic creativity. Unless we can successfully bring the arts and sciences together, we face a bleak future of either scientific advance devoid of human values or artistic innovation that is scientifically unsound. Neither offers a sustainable future.

The traditional role of the sciences has been to try to understand the complexity of the natural world. Science observes the world, develops theoretical explanations or systems of classification, then refines those theories or systems through further observation or experiment. This process has gradually replaced ignorance and superstition with understandings and explanatory frameworks. The Earth sciences began by observing and classifying rocks, but now offer explanations for the changing Earth on which we live; the formation of mountains by volcanic activity or the folding of sedimentary rock layers by the mighty forces of colliding plates; shaping of rocks by ice sheets; weathering by wind and water; carving of valleys by rivers; the development of deltas and the compacting of sediments on the ocean floor; and so on. In similar terms the biological sciences began by observing and classifying some of the millions of other species on this planet, then offered tentative explanations for the observed diversity of species in terms of competitive selection and evolution, and now gives us not only understandings at the level of the genetic code but the related capacity to produce new organisms. The health sciences began by observing illness and trying to infer patterns, then made the significant advance of identifying micro-organisms that carry infectious diseases and developing countermeasures, and now offer not only cures for many serious conditions but understanding at the genetic level of the factors influencing predisposition to those conditions.

Two broad trends can be identified in the progress of science in these areas. First, the sciences have gradually removed humans from their pedestal as God's own perfect image, and we now see ourselves as the top of the food chain, with a range of genetically influenced characteristics which limit our capacity to survive and flourish in that natural environment. Secondly, the sciences have evolved from observing the natural world to seeking an understanding of the underlying mechanisms driving the features we observe, from which we have developed the capacity to change the world rather than being its impotent observers. Thus we can now influence our own health for good or ill by conscious choices. More fundamentally, humans are an active geological agent, reclaiming ocean land or damming rivers, and an active biological agent, introducing new species into complex systems as ornaments or food producers, now even with the capacity to introduce species that have never previously existed in the form of genetically engineered organisms. So the sciences have given us the capacity to understand the natural world and that understanding has also given us the ability to change that world to suit our needs. I will return to this point below.

As well as refining our knowledge of natural systems, science has given us new ways of seeing the world. This is true both literally and figuratively. In the literal sense, science has supplemented our natural capacity to observe the world through sight, sound and touch with instruments that allow us to see the incredibly small world of atoms or the inconceivably distant galaxies, that augment visible light as the means of seeing with X-rays, infrared radiation, ultrasound, radio waves or even electrons, that allow us to see back in time to the origins of the universe or within solid materials to see the configuration of atoms that explains their properties. So we can literally see the world around us in new ways, through different eyes. Science has also given us new metaphors or concepts that influence the way we see the world. We might now see particles as bundles of energy or even probability distributions rather than miniature billiard balls, or talk about the quantum leap from one stable state to another. Concepts like relativity and uncertainty are part of everyday language, as are evolution, black holes and chaos. Thus science has not only allowed us to 'see' aspects of the real world that were previously invisible, but has also given us the ability to see the world through new lenses of metaphor and conceptual framework.

Science also provides us, as discussed briefly above, with the knowledge that has underpinned our development. What we see as the progression from the Stone Age to the Bronze Age and the Iron Age represented steps in the harnessing of scientific knowledge to improve the human condition. As this example reminds us, we naturally characterise previous historical epochs by the defining technology of the time. Our life in a modern industrialised society is entirely dependent on the products of scientific advance. The major difference between our lives and those of our grandparents at a similar age is the extent to which we have harnessed fuel energy to supplement or replace our own physical capacities. We first used the energy of burning wood for warmth and cooking, then tapped the energy of running water for the first mills. But the subsequent invention and refinement of such energy conversion devices as the steam engine, the turbine, the internal combustion engine and the jet engine have literally revolutionised life. The rate at which Australians now use fuel energy is equivalent to about 6 kilowatts per person – about the same as a chariot with four horses at a full gallop, or a small car travelling at about 50 kilometres per hour. It is roughly the same as each of us having the services of 50 slaves working around the clock in shifts, and it does for us more or less what slaves did for feudal despots: it carries us about; cooks our food; provides heating when we are cold; fans us when we are hot; provides hot water when we want to wash; constructs our buildings; makes our clothes and subsequently cleans them; provides illumination after dark; and so on. We now take for granted that we can fly interstate for conferences, exhibitions or family reunions, that we can talk to relatives or colleagues in other parts of the world, that we can watch international events unfolding as they happen on television sets in our homes, and that we can eat nutritious food regardless of the season. These are all minor miracles produced by recent advances in science and technology. The average life expectancy has been significantly increased by the almost complete elimination of infectious diseases, the dramatic reduction in maternal and infant mortality, and the recent development making it often possible now to treat conditions that were previously always fatal, such as cancer and heart disease. What were once exotic items have become commonplace. In the year I was born, the head of IBM estimated the eventual world market for computers as about five! Perhaps the most striking example of the integration of recent scientific advances into lifestyle is the portable CD player, now a standard accessory for teenagers, based on the laser and solid state electronics, which were both invented in my lifetime.

Our scientific knowledge is impressive, but it is far from comprehensive. Indeed, it has been described as islands of understanding in oceans of ignorance. Despite the valiant work of advancing science, which could be regarded in those terms as land reclamation, there is no prospect of filling in the oceans. Our ignorance of natural systems is still daunting. As an extreme example, it has been claimed that it will take about a thousand years to identify all the other species living in Australia, at the current rate of progress. So we still know only a minority of the species in natural systems, with a limited grasp of the way those species interact. We are gradually refining our understanding of complex systems, but the tendency for knowledge to be confined within disciplinary silos has led to serious problems in applying that knowledge to the complexity of the real world. The 2000 workshop on sustainability science, held in the Swedish centre of Friibergh, found that many of our most serious environmental problems are the direct result of applying narrow knowledge to complex systems. We are almost certainly still making similar mistakes through ignorance. If we want to be confident we are meeting human needs without damaging natural systems, we will have to make considerable advances in our understanding of those systems.

Just as the sciences have broad roles in our society, so have the arts. The first of these could be called 'representational'. The arts show us who we are, where we are and what we are. The painting, the music, the film and the literature of Australia are uniquely ours. Nobody else paints our landscapes, or writes plays that depict our society, or creates music based on our unique identity. Our arts have clear antecedents in other cultures, ranging from those of Western Europe to those of the Indigenous people, but are also increasingly influenced by those of Asia and the Pacific Rim. They are constantly changing and evolving as our society changes and evolves; the films and the music and the poetry of 2002 are quite different from those of 1952. The techniques of representation are also constantly changing, with inevitable consequences for the messages we receive. In the words of Marshall McLuhan, the medium is the massage. Even such a small change as using videotape rather than film to tell a story inevitably massages that story, giving it a different slant. An insensitive filming of a book, such as the Hollywood treatment of *Captain Correlli's Mandolin*, produces something completely different from the original. But even a determined attempt to stick closely to the book, such as *Harry Potter and the Philosopher's Stone*, still gives an altered reading of the original text; the text has inevitably been massaged by the choice of medium.

The differences are also influenced by the individual lenses of culture and history through which we experience the art form. No two individuals will perceive a sculpture, a film or a musical composition in the same way. It is important to have Australian artists and media institutions to portray us to ourselves, but Australia is not homogeneous. Differences of age, of gender, of social class and acculturation will cause different artists to approach their subject in different ways, and will similarly cause each of us to respond in our own unique individual way to the finished work.

The arts are often 'aspirational'; they go beyond a portrayal of who and what we are to reflect what we could become. The arts constantly remind us that there is not one fixed, predetermined future toward which we are inexorably marching, step by daily step. There is always a range of possible futures, some of which are more probable than others and some of which appear preferable to others. Which of the range of possible futures actually occurs will be the product of our decisions and actions, individual and collective. Just as the Australia of today has been shaped by the choices of successive generations of Australians, so we are all engaged in writing the book of the future. The arts constantly encourage us to grapple with this responsibility. One stream of exploration of possible futures contains the dystopias like *1984*, *Brave New World*, the *Mad Max* movies or *The Matrix*. By their graphic portrayals of possible futures, they warn us of the consequences of some likely developments. AD Hope's poem 'Toast for a Golden Age' told us 40 years ago that greed, insensitivity and the growing human population would inevitably damage natural systems and imperil biological diversity. Ross Edwards's sublime musical composition performed in Sydney at the dawn of the new millennium was an inspiring evocation of the harmonious and culturally diverse Australia we could become, while the writings of Donald Horne, Bob Ellis and Phillip Adams remind us of the alternative future to which we could be taken by the bigoted myopia of John Howard.

Finally, at their best the arts are 'inspirational', elevating us to seek a better tomorrow. It can be a new performance of a traditional art form, as in refreshing local interpretations of European operas or Shakespeare's plays. I doubt if anyone who experienced Adelaide's staging of Wagner's Ring cycle will ever forget either the production or their pride that it had been done here. It can be an innovative new look at a traditional form, perhaps a relocation like Baz Luhrmann's *Romeo + Juliet*, or a radical re-interpretation like Robyn Archer's *Boy Hamlet*. It might be a poem, a novel or an essay, a painting or a sculpture, a musical composition or an innovative drama. We might witness it performed, see it in a gallery or read it on the train. All of us can remember moments when the arts have transcended the representational and the aspirational to become truly inspirational, giving us simultaneously a vision of a better world and the motivation to work toward it.

So how do the arts interact with the sciences? Science affects the arts, as is abundantly clear from the works of art which form the exhibition component of *conVerge*. But the arts also affect science. Most importantly, the only real hope for a sustainable future involves a creative synthesis of the arts and the sciences to develop new ways of meeting our needs and our hopes. It is no exaggeration to say that the future of our country depends on our ability to meet that challenge.

Science influences the arts by giving the artist new conceptual frameworks, new ways of seeing the world. Sometimes the influence is obvious, as when our knowledge of the solar system inspired Gustav

Holst to compose his symphonic suite *The Planets*. At other times the influence is more subtle, as in the way Bertold Brecht used the life and work of Galileo Galilei to make more general observations about freedom of inquiry and the suppression of dissent, or Michael Frayn used the 1940s discussions of quantum mechanics to ask broader questions about the responsibility of scientists in his new play *Copenhagen*, to be staged in Adelaide later this year. Aldous Huxley and George Orwell used what were, when they wrote, conjectured possible advances in genetics and information technology to make similar points in *Brave New World* and *1984* respectively. The vision of the Earth from space, first available to us less than 40 years ago, has inspired generations of artists. The same point could be made about either large-scale astronomical objects such as galaxies or the infinitesimally small, such as assemblages of atoms; the new images inevitably act as a stimulus to creativity. Several of the works of art forming the exhibition component of *conVerge* are clearly inspired by such recent scientific advances as the human genome project, DNA sequencing, ultrasound imaging or cell cultures. A scientific advance can almost amount to an art form in itself, as in the glorious Mandelbrot images which are the most public manifestation of chaos theory, or David Malin's spectacular pictures of the night sky.

So science can widen the opportunities available to the artist by providing new conceptual tools, or new ways of seeing the world. It can also provide new techniques, or new ways of achieving traditional goals. New materials can give the visual artist or the sculptor new opportunities: acrylic paints; durable alloys; resin-bonded compounds. New techniques can make it easier to achieve traditional forms, as with the use of computers to generate a mould for casting a sculpture, or can become a new art form in itself, as when the computer becomes the design tool to create a novel shape or form for the sculptor. The advance can enable a new form of artistic expression. Laser beams are a good example of a scientific advance that can either be an independent art form, where the light show is the creative work, or can augment another art form; I will never forget the use of red laser beams in the closing moments of Adelaide's 2001 staging of *Parsifal*. There are several examples in the exhibition of scientific imaging techniques being used in works of art. New technology can even gradually transform an established art form. My favourite example is the way improved techniques for constructing and maintaining musical instruments have made it possible for performers to gradually tune to a higher pitch, giving a more brilliant sound. The pitch of the note A, to which orchestras tune, has gradually increased from about 420 Hertz at the time of Mozart to 440 today. The change makes no difference to orchestral players, but it has a serious effect on singers, who must produce the same pitch as the musical accompaniment. The famous note of high C, seen as the Mount Everest for opera singers, was more like a modern B flat at the time most operas were written, a note within the reach of most chorus tenors and sopranos.

Equally significantly, the arts have influenced science. When I was an undergraduate, it was common to portray science as socially and politically neutral, devoid of emotional or moral content. This widespread view was spurious, as the great creative scientists always put considerable store on their instinct or the aesthetic appeal of the theoretical models they derived, thus celebrating the emotional dimension of their science. The arts have now taught the sciences more generally, though not yet universally, that our perception of the world is inevitably influenced by our values, our culture and our predisposition – that our engagement with nature and the production of knowledge is invariably a social process. The arts have also taught us that the application of science to technology is always a political act. We can never change only one thing in a complex system; the use of technology always produces losers as well as winners, costs as well as benefits. So there is a responsibility on scientists and technologists to be aware of the consequences of their work. Some knowledge is neutral and can be used for good or ill. The laser has no political content in itself, but there is enormous political significance in the choice to use it for entertainment, for healing or to guide weapons of mass destruction. Other knowledge has embodied political content: the neutron bomb or biological weapons can only be used to kill people. An alarming fraction of the world's scientists are still engaged in military research, aimed explicitly at finding cleverer ways to kill other people, or to find clever ways to stop other clever people from killing us. That is an absurd misuse of limited resources.

As I said in the introduction to this essay, the urgent task for human society is to find ways of meeting the growing aspirations of an increasing population on this finite planet without compromising the natural systems on which we depend. That will require increased scientific understanding, but it will require that understanding to be driven by our humanity rather than our greed or our aggression. It has been said that we are not passengers on spaceship Earth but the crew, and that we need to take our responsibilities seriously. We are the crew because our scientific knowledge gives us a better understanding than we have ever had before of the natural systems of this planet. We are also the crew because our technology gives us more options than we have ever had for meeting our material needs within the limits of those natural systems. But most basically, we are the crew because our humanity demands that we use that scientific knowledge and that technological capacity wisely in the interests of all humans, now and in future generations, and all other species. The arts do us a huge service in constantly reminding scientists and technologists of that responsibility. The only hope for a sustainable future lies in a creative fusion of our scientific understanding, our technological capacity and the insights into the human condition provided by the arts.

CHANGING PUBLIC ATTITUDES TO LONG-TERM ISSUES

Edited extract from article for Griffith Review 12, *2006*

If industrial society is to survive, the next century will have to be a time of transformation, not just in technological capacity but also in our approach to the natural world, and to each other. The second report in the UNEP series on the global environmental outlook said: 'The present approach is not sustainable. Doing nothing is no longer an option.' A sustainable society would not be eroding its resource base, causing serious environmental damage or producing unacceptable social problems. Our present lifestyle does not satisfy *any* of those main criteria. We are dissipating resources that future generations will need, damaging environmental systems, and reducing social stability by widening the gap between rich and poor. It is possible to move to a sustainable future, but it will require fundamental changes to our values and social institutions. While UNEP said that doing nothing about the huge problems we face is not an option, it remains the most common response of today's decision makers; our national strategy, agreed in 1992, gathers dust in government pigeonholes. So how likely is it that we can achieve the fundamental changes needed for our civilisation to survive? While sceptics point to the durability of cultural institutions and the reluctance of people to make voluntary sacrifices for the common good, we know that cultural traditions do change and people do make voluntary sacrifices. I explore below examples of changes and illustrations of the willingness of people to make short-term concessions for the long-term good, or to accept restrictions of their personal freedom for the good of the whole community. These examples suggest practical ways of achieving the sort of transition we will need for a sustainable future.

Perhaps the most dramatic example of recent change in our culture is the shift in the last 40 years of attitudes to smoking tobacco, especially in shared spaces. In 1972, a persistent writer of letters to the existing domestic airlines celebrated the concession made by one of them to introduce a limited trial of setting aside a few rows of seats on some flights to be designated 'non-smoking'. Up until that point, smokers presumed they had the right to light up anywhere on aircraft, buses and trams; the same tolerance applied to restaurants, meeting rooms, public buildings and, in some states, even to cinemas. The change on airlines was truly dramatic. With a choice, more and more customers asked for 'non-smoking' seats until finally, in 1992, the Australian Government announced a total smoking ban on domestic flights. There was barely a flicker of protest against the total prohibition of an activity that had been presumed acceptable only 20 years earlier. In the workplace, the crucial event was the successful legal action by Liesel Scholem against the New South Wales Department of Health. Ms Scholem presented medical evidence to show that her emphysema was a consequence of the department's allowing smokers to pollute the air she breathed in a shared office. Within a year of the verdict, many workplaces became non-smoking areas, forcing smokers to stand on the footpath outside their buildings. With the evidence accumulating that 'passive smoking' is a health hazard, the prohibition has spread to most public transport vehicles, public buildings, restaurants and even pubs. Most smokers have conceded that their right to smoke is overruled by the rights of others not to breathe exhaled tobacco smoke. As the right to smoke has been curtailed and more people have become aware of the health risks of smoking, the percentage of the adult population who smoke tobacco has fallen dramatically, from about 50 per cent in 1950 to below 20 per cent today. That is a radical change, by any standard.

There are other, equally striking, examples of radical changes in attitudes. When I first obtained my driving licence, police were allowed to stop a vehicle if they had reasonable grounds for believing the driver could be drunk. So a driver who attracted attention, for example by weaving all over the road or going through a red light, could be stopped, but an inconspicuous drunk could drive with impunity. Random breath tests were widely considered an infringement of personal freedom when first suggested in the 1960s. We are now accustomed to a regime in which drivers can be stopped 'anywhere, any time' and invited to provide a sample of their breath. There is a close parallel with seatbelts in cars and helmets for motorbike riders. In each case, the compulsion was seen by some as an unacceptable intrusion, despite the overwhelming statistical evidence that the measures saved lives. Australia introduced compulsory seatbelts before most other nations; I recall bemused critics in the northern hemisphere seeing the measure as evidence we were sliding down the slippery slope to some sort of police state!

When I was young, plastic shopping bags did not exist and shoppers took their own permanent bags to shops. The plastic bag arrived and spread, so for a few decades there were very few shoppers taking their own bags to the supermarket. Now that increasing numbers of people are concerned about the wasteful practice of using throwaway bags, we are seeing a return to the practice of shoppers bringing their permanent shopping bags. In some suburbs, those still using plastic bags are a small minority, attracting the attention and sometimes the opprobrium of other shoppers. As a final local example, in the 1950s soft-drink bottles incurred a deposit, so users returned them to get a refund. There was no deposit on beer bottles, but community groups like the Scouts collected them and were paid to return the bottles to distributors. We then passed through a throwaway era, with bottles used once and put into rubbish bins – a radical change. Recycling began in a half-hearted way, requiring trips to designated collection points, so few people took the trouble to return empty bottles. Then kerb-side recycling was introduced in many areas, with a consequent dramatic increase in the amount of glass, metal and plastic recycled. The state of South Australia has re-introduced deposits on beverage containers. This measure has been very successful in boosting the level of recycling. In a direct response, the packaging industry has lobbied feverishly to prevent this practice spreading! As with shopping bags, we passed through a stage of wasteful resource use before returning to the more responsible use pattern of 50 years ago: two opposite radical changes in less than the average human lifetime.

Probably the most striking example of radical change in the last 20 years is the end of communism, most dramatically symbolised by the fall of the Berlin Wall. For decades the wall divided the city of Berlin, patrolled by armed guards to prevent people fleeing gloomy East Berlin for the bright lights of the West. As a physical structure and as a symbol of the power of the Communist government of the DDR, it seemed absolutely impregnable. Only a handful of people succeeded in escaping to the West; many perished in the attempt. Then, almost literally overnight, the Wall was abandoned. Joyful Berlin residents tore pieces out with their bare hands. Today Berlin is a united city, once again the cultural and political centre of Germany. A small museum has preserved the history of the wall and entrepreneurs sell small fragments of the structure to tourists, but for most visitors and residents, it is as if the barrier had never been there. The waters of history have closed over it.

The end of apartheid in South Africa is a similar story. For nearly a century after the Boer War, the Afrikaners ruled the country with their strange mix of Christianity and racism. The people of South Africa were divided into the three groups of whites, 'coloureds' and

Africans, a process that required some arbitrary decisions despite penal sanctions for sexual activity between members of different classifications. Public spaces like beaches and even park benches were marked according to which group could use them. Those seeking change inside the country, like Nelson Mandela, were imprisoned or executed. Internal protests were ruthlessly suppressed in such actions as the infamous Sharpeville massacre. Anti-apartheid feelings in other countries became stronger, but conservative politicians sided with the regime, with the Queensland premier in 1970 going to the ridiculous lengths of declaring a state of emergency to allow police to deal with those protesting against a rugby tour. Sports people who refused to play against official South African teams were ostracised, in some cases ending their representative careers. Calls for economic sanctions or sporting boycotts were derided by business leaders, sporting officials and most politicians. Eventually, the rising tide of political reaction forced politicians in many Western countries to support pressure on the South African regime, leading in time to the dismantling of apartheid and the enfranchisement of the entire adult population. Nelson Mandela, released and almost saintly in his willingness to forgive his tormentors, became president of the 'rainbow nation' and the long-banned 'Nkosi Sikelele Africa' was incorporated into the national anthem, with Mandela's wearing of the country's rugby jersey at the World Cup final a powerful symbol of reconciliation. Many of those unable to face the dismantling of their system of privilege fled the country but most remained and adapted to the new reality of 21st-century South Africa. While a hundred years of institutionalised disadvantage meant there was a need for some arbitrary measures to accelerate change, such as quotas in sporting teams, South African cricket and rugby teams are increasingly representative of the diversity of the country. As with the Berlin Wall, it is difficult now for a younger generation to imagine the old system even being defensible, let alone appearing unassailable.

There are more modest examples of radical changes in community attitudes enabling political reform. London Lord Mayor

Ken Livingstone decided to try to cut through the policy impasse that was making the city unworkable. With the public transport system becoming steadily older and less attractive to commuters, more people were taking cars into the city, producing congestion and pollution. The reduced patronage of the public transport system meant it was unable to raise the revenue to invest in improvements. Livingstone implemented a 'congestion charge' – essentially a levy on those bringing motor vehicles into the inner city area – to fund the revitalisation of public transport, arguing that sticks (financial penalties) and carrots (better public transport) were needed to change behaviour. The levy was not trivial, equivalent to about A$12, and there were dire predictions of economic and social chaos before the introduction of the scheme. It was so successful that after the first year, Livingstone had widespread support for increasing the levy to the equivalent of about A$20 to speed up the raising of funds to improve the bus and underground rail system! Far from creating chaos, the change has made the inner city much more pleasant – the streets are less congested and the air is cleaner, leading to an increased willingness of people to be in the city. Instead of the feared collapse of commerce, shops are doing more business as people return to the centre of the city. This was confidently expected by the planners, because it is exactly what has happened when streets were closed to create shopping malls, but there was still widespread opposition before the change. Now that the change has been made, it would be politically very difficult to return to the old ways.

I have argued earlier (see page 165) that there are four steps to major change: discontent, a new vision, viable pathways and commitment. Unless there is discontent, there is no motivation to change. If I am happy with my fitness, I am unlikely to change. If we are happy with our lifestyle, we are unlikely to change – which is why advertising tries to make us discontented and therefore more susceptible to the latest consumer fad. Richard Eckersley argues that the old seven deadly sins – envy, greed, lust, pride, laziness and so on – have been repackaged as the seven marketing imperatives of the modern world! They are constantly used, as Clive Hamilton put it, to urge us to spend money we don't have to buy things we don't need to impress people we don't like. Producing dissatisfaction is the aim of all advertising. I am dissatisfied with our lifestyle, because we are depriving future Australians of the sorts of opportunities we have taken for granted. Unless that awareness is widespread, there will not be the dissatisfaction that will motivate change.

Being discontented is not enough. Even if I am unhappy with my fitness or my lifestyle, I won't do anything unless I can imagine a better alternative. Without a vision, change might actually make things worse rather than better. So the second step to real change is to picture the future we want. Advertisers use this principle, portraying the life of indolence or the sexual pleasures that await us if we buy their product. In the case of my fitness, I can imagine the sort of state I would like to achieve, because I was there a few decades ago! Achieving a sustainable society is more difficult because we haven't been there, so we have to invent it. Unless we have a credible vision of a future sustainable society, it is very unlikely there will be support for change. In writing *A Big Fix*, I tried to describe the essential components of a future sustainable society. There is room for argument about some of the details, but the general principle is clear. We must limit human consumption so it doesn't exceed the sustainable level of production from natural systems.

Again, even a coherent vision is not enough if we can't see a way of getting there from where we are. I do a fair bit of bushwalking, so I know that the easiest route from your starting point to your destination is hardly ever a straight line; you need to take account of the lie of the land and the vegetation. A friend might advise me against trying to recover my fitness of 20 years ago from where I am, but I might see a viable pathway, with suitable attention to diet and exercise. So the third component of change is developing and describing feasible pathways that would take us to our goal from where we are now. Again, it is more difficult with a future sustainable society. Even when a future sustainable society can be imagined and described, it is difficult – and contentious – to spell out the steps that will take us there. Since a sustainable future must involve much less use of fossil fuels like coal and oil, the initial steps run counter to some very powerful vested interests. In Australia's case, the export coal industry and the heavily subsidised aluminium industry have been remarkably effective in blocking the sorts of changes that are clearly in the public interest. Before the1997 Kyoto conference on climate change, which set preliminary targets for reducing the levels of greenhouse pollution from the industrialised world, I attended a briefing for interest groups. I sat with mounting incredulity as the leader of the Australian delegation set out the stance being proposed for the Kyoto meeting. When I asked whether it was possible to identify any nuance in the official position that differentiated it from the public stance of the energy-intensive industries, the delegation leader responded angrily, but was eventually forced to concede that there was no significant difference! The position adopted at Kyoto by the Australian Government, led by a member of the Howard Government's ministry, was to defend the interests of two industries that are largely overseas owned, the export coal and aluminium businesses. Another example of a vested interest that has been very successful in blocking change is the road freight operation, heavily subsidised because the registration charges and fuel taxes paid by large trucks are tens of thousands of dollars less than the damage they do to roads. These examples illustrate that the pathways to a sustainable future must be politically feasible as well as technically viable.

The fourth component is commitment. If I can spell out a coherent program of diet and exercise that will recover my physical fitness, it will still take commitment to stick to the regime. We all know people who have good intentions, but lack the will to stick to their plan. The politics of sustainability hinges on developing such a public commitment to a sustainable future that the change becomes politically irresistible. As the examples above show, traditional politics will only respond to these urgent long-term issues when there is irresistible pressure. The historian Paul Kennedy argued that those who succeed in democratic political systems are usually those who have refined to an art form avoiding any threat to powerful interest groups. So they are unlikely, he proposed, to adopt significant changes in the interest of future generations as long as they can argue that experts are divided or more research is needed. In the case of complex issues like global climate change, there will always be some division among experts, making it feasible to claim that more research is needed before we embrace significant change. It has become steadily more difficult to find scientists with any credible claim to expertise who will deny that human activity is changing the global climate, but media organisations and governments have responded by intensifying the search rather than accepting the overwhelming consensus! While several European countries have now adopted serious targets for reducing their greenhouse pollution, in Australia we are still getting evasion and spin from our national government. The 2008 proposal for a Carbon Pollution Reduction Scheme (CPRS) involved so many hand-outs to big polluters that it might have been designed to ensure no significant change happens.

All around the world, people are striving to develop social and institutional responses that will enable the transition to a sustainable future. We have to recognise above all else that we share the Earth with all other species and hold it in trust for all future generations. That moral duty provides the impetus for the transition to a sustainable future, just as moral principles underpinned the end of apartheid and the dismantling of the Berlin Wall. As in those cases, the changes we need are not inevitable, but those examples serve as a reminder that radical changes happen when enough of us want them.

A CASE FOR CHANGING AUSTRALIA DAY

The Weekend Australian, 26–27 January 1991

This weekend we celebrate Australia Day on 26 January. It commemorates the arrival at Sydney in 1788 of the first Europeans who settled here. Landing in the blazing heat of a Sydney summer after a nightmare journey from their cold, damp European homelands, it is not surprising that they saw this as a bright and savage land. They found a climate and soils totally alien to them. Too proud to accept the advice of those who already lived here, they did their best to transplant their agriculture and what they knew as civilisation.

I dislike some elements of our European inheritance: the anachronistic retention of allegiance to the British monarchy; the widespread imitation of European dress styles; and the colonial relic of the Union Jack on our flag. There is also much for which we can be grateful in our institutions derived from Britain. But 26 January should be a day for deeper reflection on our culture. Part of that reflection should be its reminder of the need to forge a proper peace with the

Aboriginal people, whose land we took by a process that can only be described as robbery with violence.

I grew up with the comfortable myth that the Aborigines were nomads who cheerfully and peacefully retreated from the lands wanted by the European invaders. Recent work by such scholars as Professor Henry Reynolds has documented the long and bloody guerilla war which actually took place. Though they were using spears and boomerangs against guns, the Aborigines fought tenaciously to retain their traditional lands.

I recently attended a conference in Perth. One of the highlights was a presentation by Professor Fay Gale, Vice-Chancellor of the University of Western Australia. Now that most vice-chancellors are full-time bureaucrats, finding one of them still pursuing scholarly work is itself a joy to behold. More importantly, I found myself spellbound by Professor Gale's account of her study of pre-European Aborigines. Most of us know relatively little about those people, who were at least as different from their modern descendants as European Australians are from the wretched inmates of the convict ships.

Some desert tribes were largely nomadic, but most of the Aborigines lived as settled communities along the coastal fringe of the continent. Studies of the languages show that each community occupied a defined area, usually separated from neighbouring groups by mutually recognised boundaries. While much of the food eaten by Aborigines was collected from the bush, they also planted and harvested. Indeed, there is evidence that they used irrigation systems, especially in Arnhem Land and the Grampians. They certainly used fire to manage the natural environment for food production. Some of the indigenous foods were of high quality. The cereal crops used to make bread had more than twice the protein content of wheat or rye. Root crops were much richer in vitamins and minerals than potatoes; some native fruits were exceptionally rich in vitamins.

Our early forebears failed to learn appropriate food production techniques from the Aborigines. They nearly starved as a result. Burke and Wills perished in a place where no Aborigine would have wanted for food or water. The ABC television programs on bush tucker have brought to an urban audience some of the sorts of knowledge which every Aborigine needed.

As would be expected of settled communities occupying areas with different productive capacities, the pre-European Aborigines had an extensive network of trade routes spanning most of central Australia. They had a style of mapping which was certainly not the same as our Cartesian system. While it seems primitive to us and would not be useful for such tasks as finding a suburban shop, it was ideal for finding waterholes in the desert. It enabled Aborigines to accomplish, quite routinely, navigational tasks which defeated our European forebears.

Professor Gale noted that the Aboriginal women of last century had much more independence than European women. They were economically independent, had a separate legal system and went away to women's camps. Compare that with the life of a white Australian woman in 1890. While we tend to regard Aboriginal culture with a Eurocentric condescension, its various facets – art, music, dance and story – all played an important role in the enduring relationship between the people and their environment.

This suggests a compelling reason for making a proper peace with the descendants of the Aborigines we displaced from most of this country. We should be trying to learn from them techniques for living sustainably. After 200 years of European settlement we are now aware that many of our practices cannot be sustained. We have set up study groups to establish principles for ecologically sustainable development because our lifestyle is so far removed from ideal. Our science is not yet sufficiently precise to tell us how long Aborigines lived here before 1788, but it appears to have been at least 40,000 years. That is a remarkable record of survival for their culture.

Throughout that period, Aboriginal culture changed and modified the natural environment. It also, however, embodied principles and practices for retaining the productivity of natural systems. We really need to learn those principles and develop those practices.

When we make our peace with the Aborigines, we might be able to take up the suggestion endorsed by DD McNicoll in *The Weekend Australian* of 5–6 January. Since the date of 26 January means nothing to many Australians and is an offence to others, we could move Australia Day to commemorate a date which is really worth celebrating. That is the anniversary of 1 January 1901, when six feuding colonies became a fledgling nation. At least then we could guarantee that most Australians would celebrate the national day.

LIFE'S BUT A SHORT SPAN FOR MAN

The Weekend Australian, 9–10 March 1991

A typical Australian family now spreads over several states. Brothers and sisters only get together for family weddings, funerals and other rare events. A recent family wedding in a small country town reminded me these rituals have gradually changed over the years. The marriage vows are now more egalitarian, more about loving and sharing, less about honour and obedience. These days the ceremony is less likely to have a religious emphasis. The traditional telegrams are no longer physically delivered by young messengers on bicycles; they are also rather more explicit than the time-honoured coy allusions to troubles only being little ones.

Some aspects are unchanged. The same toasts are proposed, bridal bouquets are still thrown and every effort is made to spoil the happy couple's first night of connubial bliss. At the wedding I attended, we realised that the bride and groom between them had four grandmothers, but no surviving grandfathers. It was a gloomy reminder of the health penalty of being born male.

Men are, on average, physically stronger than women. We also tend to be more aggressive. These characteristics distract attention from the fact that we have a shorter average life expectancy. In that sense, men are clearly the weaker sex. Our survival chances are worse all along the line. The male foetus is more vulnerable than the female, male infants have a 20 per cent higher mortality than females, young boys have more fatal accidents than young girls, young men are more likely to die in road accidents than young women and middle-aged men are more prone to fatal heart attacks than women. Overall, the average life expectancy of women is about eight years more than the figure for men. Moreover, the difference has been growing for a hundred years. The 1986 census found that women account for 73 per cent of the people over 85 and more than 80 per cent of those over the age of 100.

There are some complications arising from recent social changes. As one example, men are much more likely than women to suffer from lung cancer or ischaemic heart disease. One of the reasons is that many more men than women have been tobacco smokers. There has been a significant decline in recent years in the level of smoking by men. At the same time, there has been a successful campaign to recruit young women into the habit, and they now smoke in unprecedented numbers. It is a gloomy certainty that there will be a corresponding increase in the number of women who contract lung cancer and other health problems such as chronic bronchitis, since there is a clear link between smoking and these diseases.

I recently received a letter from a lawyer who claimed never to have seen a death certificate giving smoking as the cause of death. This may be a true statement, but it is the sort of sophistry which gives the legal profession a bad name. It is true that death certificates are unlikely to give smoking as a cause. It is equally true that there is a clear link between smoking and lung cancer, chronic bronchitis and heart disease, all of which are principal causes of death.

More women are now in the sort of stressful occupations that have traditionally been associated with hypertension and related health problems among men, so we may see a growing tendency for these problems to affect women. While women live longer than men, they have their own health problems. Pregnancy, abortion and childbirth, while much safer today than in our grandparents' day, are still significant health hazards for women. Some cancers affect only women: cancers of the cervix and uterus are obvious examples. These diseases have their own distinctive risk factors.

The probability of women getting breast cancer, for example, is influenced by child bearing. Relative to childless women, those who have their first child before the age of 20 are only half as likely to get breast cancer. On the other hand, a woman who has her first child at the age of 35 has a 20 per cent greater risk than a childless woman. The differences are believed to be due to changes in breast tissue during pregnancy. While this knowledge might be seen as an incentive for early reproduction, the world is not that simple. The risk of cancer of the cervix is increased by early sexual activity, so contemporary society's acceptance of early sexual experiences may be one reason why rates of cervical cancer among young women have doubled in the past 20 years. Looking at these two risk factors suggests that the best health strategy for a woman might be an early pregnancy without sexual contact, while the more common practice of early sexual experiences without getting pregnant carries the sort of health penalty to bring a smile to the lips of Fred Nile.

Women suffer more chronic illness and disability than men. As a consequence, they are more likely to use drugs such as tranquillisers, pain-killers and sedatives. In their book about lifestyle and health, *The LS Factor*, Professor Tony McMichael and Dr Basil Hetzel associate the greater incidence of depression and anxiety among women with their lack of economic and social power. This is a reminder that health is not simply associated with physical factors. Health is not just an absence of illness; a healthy individual has a general feeling of well-being. Thus you are more likely to be healthy if you believe in your own worth and your capacity to control your own future. Those who are oppressed by others are less likely to be healthy. This is true whether the oppressor is Saddam Hussein, General Pinochet or your own family.

Even if I am not oppressed by anyone, I still suffer from the basic biological limitation of being male. As I was reminded last week, there's not much chance I will dance at the weddings of my grandchildren.

AD MEN SHOULD EAT THEIR WORDS

The Weekend Australian, 13–14 October 1990

The campaign to restrict advertising of junk food to children has evoked an amazing response from the industry. The Australian Consumers' Association (ACA) has launched the push to control TV advertising aimed at children. An open letter to the prime minister and statements in the press attacked the way the ads promote consumption of foods high in fat, salt, sugar and total kilojoules.

There is no doubt about the emphasis of advertising directed at young people. ACA surveyed a week's advertising in the 3.30–6 pm time bracket on Sydney commercial TV. More than 30 per cent of all ads were for food products. The report in the October issue of *CHOICE* said that 75 per cent of those food ads were for 'foods like chocolate, lollies, biscuits, soft drinks, sugary cereals, fries and other fatty foods'. A similar survey was conducted in Adelaide by Heather Morton. This showed an even greater number of food ads, making up over three-quarters of all commercials between 4 pm and 5 pm, with the figure for one channel being 93 per cent. As in Sydney, most of the ads were for heavily sugared drinks and cereals, chocolate and 'fast food' restaurant chains.

The reason for concern is that the foods advertised should only play a small part in a properly balanced diet. This is agreed by nutritionists quite generally. There is particular concern in Australia about our eating habits; despite a reduction in the consumption of some problem foods in recent years, many of us still have unhealthy diets. The consequences of poor diet are clearly visible in the soaring health budget. The recommended dietary guidelines released by the Department of Community Services and Health suggest we should drink less alcohol and eat less fat, sugar and salt. We should eat much more fresh fruit, vegetables and cereals, preferably wholegrain; indeed, we should probably aim to get the bulk of our nutrients from these sorts of foods.

The stream of TV advertising runs directly against those guidelines. The constant message to young children encourages them to consume foods which should only constitute a small fraction of their diet. As *CHOICE* points out, concern about food advertising directed at children is nothing new. The National Health and Medical Research Council drew attention to this problem in a 1981 report, urging the food industry to implement self-regulation. As often happens, there was a short period of restraint after the report was published, but the problem is now even worse than prompted the call for action ten years ago.

The 1981 approach was to urge the industry to be responsible. In the current climate of deregulation, the modern euphemism for abdication of government responsibility, it is hard to persuade our elected leaders to intervene. However, a 1988 CSIRO survey showed more than 70 per cent of adults favoured action to restrict food advertising aimed at children.

The ACA campaign provoked an interesting reaction from the advertising industry. The director of the Advertising Federation of Australia, Mr Bruce Cormack, angrily accused the consumer movement of being in favour of bans and restrictions which are 'inevitably counterproductive'. If restrictions really are counterproductive, presumably the industry will support them as a cheap way of increasing demand. Indeed, if advertising bans increase sales as well as saving money, I can only wonder why the industry isn't asking the government to help them by banning TV commercials more generally.

Mr Cormack went on to deny the ACA claim that children were likely to be influenced by TV advertising. This newspaper quoted him as saying 'substantial literature from around the world, published in authoritative journals and literature, has shown this to be wrong'. Leaving aside the interesting reflexive notion of literature which is published in literature, this still appears to be quite a remarkable claim on behalf of the advertising industry. Though believing their TV material to be ineffective in persuading children to eat junk food, they nevertheless continue to make and show advertisements with that apparent purpose.

This unselfish willingness to keep the wheels of the economy turning by funding fruitless ads clearly constitutes a substantial act of philanthropy, deserving of more widespread public recognition. Perhaps the industry could institute a new annual award for the ads which are judged to be most cost-ineffective. There are some I would like to nominate.

There is more to this question than meets the eye, however. It is not the advertising industry that funds the avowedly ineffective ads, but their clients. Has the industry revealed to its clients the ineffectiveness of its ads? This is an important issue, because we are entitled to know whether our communal gratitude should be directed at the advertising agencies or at their customers, the promoters of junk food.

I don't like to reinforce the widespread public view that academics are cynical, but I do have some difficulty with the notion of selfless denial being put forward. I know that some perverse individuals make a point of avoiding those products promoted by ads which clearly assume the average viewer has the intelligence of a wombat and the attention span of a grasshopper, but I assume those paying for the ads can do their sums. They pay for ads if they deliver extra sales.

The ACA survey found that only 4 per cent of food ads promoted the kinds of food which should predominate in our diet. The lesson of the 1980s is that self-regulation does not work. We urgently need national guidelines governing this aspect of TV advertising.

The eating habits of today's children will literally determine the shape of our future. We will have a healthier community if we encourage good diet. There should be a levy on advertisers of junk food and it should be used to promote healthy eating habits.

THE SUMMER WE SUITED OURSELVES

The Weekend Australian, 17–18 November 1990

Temperatures are rising. For most of us the long hot summer has begun. It is an appropriate time to ask why we don't dress sensibly for our climate. For more than 200 years most of us have dressed like Europeans. When the First Fleet arrived early in 1788, the British military clumped ashore in their thick woollen uniforms. They sweated uncomfortably in the blazing Sydney sun at the height of the Australian summer, watched by understandably bemused Aborigines. With the arrogance characteristic of the colonial era, the invaders thought the Indigenous people were savages because of the way they dressed.

Our national dress habits did not change for a very long time. I was told recently by one of the pioneers of the Tasmanian wine industry of his arrival in Sydney more than 50 years ago. His first impression was of the men wearing heavy, dark suits and complaining that it was 'muggy'. During the 1960s I spent some time trying to interest school students in the obscure mysteries of the science syllabus, much of which seemed to have been deliberately chosen to have the least possible relevance to their lives. I literally sweated through the Sydney summer because male teachers were required to wear ties as a mark of our professional status. How we envied the physical education teachers in their skimpy shorts. A few of us bravely wore tailored shorts and long socks but it was made clear to us we were letting the side down by this unprofessional behaviour.

We were aware things were different in Queensland; there were rumours of people getting around like tropical planters in shorts and long socks. Then Don Dunstan bravely appeared in public in shorts before the startled burghers of Adelaide; surely this was the harbinger of a new age of dressing for our hot summers? Well, perhaps it should have been but it wasn't. To this day most of our male business executives wear suits to work, even in the height of summer. The only obvious change from the 1930s is that the suits are more likely to be a light grey than a heavy blue serge. Not only do business people dress in this anachronistic way; so do politicians, many public servants, and even academics in some of our old-fashioned universities look like tailors' dummies.

Why do we do it? It is at least partly the cultural cringe. People in the business centres of the Northern Hemisphere wear suits, as we see all the time in the mindless junk which is brought to our television screens by the forces of the market. Many of us believe that we must look like European or American corporate barons to be seen as successful.

The problem is that this style of dress is just not suitable for our summers. If you try to compare the latitudes of our cities with those of Europe, you will find that most of Australia isn't equivalent to anywhere in Europe. Hobart is about the same distance from the equator as Marseilles; Melbourne is about the same as Athens. Further north we run out of European equivalents. Sydney is about the same latitude as Tripoli or Casablanca. Brisbane is about the same as Bahrain. Cairns is the same as Timbuktu. Even the British don't usually wear suits in Egypt, the Gulf or the Sahara.

If you visit Singapore, Indonesia or the Philippines, you find the men are dressed sensibly for the climate in open-neck shirts. They do so because the temperature is above 30 degrees Celsius so it is rational to dress for comfort. But the mercury regularly reaches those levels in our cities in summer. Shouldn't we dress in summer like the Filipinos or Indonesians?

The first response of many people is to see it as a joke. When I floated the idea of dressing less formally at a conference recently, one Sydney newspaper illustrated its report with a cartoon of naked businessmen, lending a whole new meaning to the idea of full disclosure. Another cartoon showed politicians wearing shorts and singlets, suggesting that dressing down would reduce the dignity of Parliament, as if anything could do that more effectively than the recent behaviour of some honourable members.

The standard of dress worries me less than shouted abuse as a substitute for reasoned debate.

One of the consequences of not wearing clothing appropriate to our climate is a massive waste of fuel energy. If you dress as you would in Europe, you need a European temperature to be comfortable enough to work effectively. Cooling a building from an outside temperature of 30 degrees to an inside level of 20 degrees takes twice as much energy as cooling it to 25 degrees. Many of our buildings are cooled so aggressively that those who dress sensibly are uncomfortably cool. I often see women, who generally dress more sensibly for our summer than men, wearing sweaters in January in air-conditioned areas. We waste enormous amounts of electricity (and millions of dollars) each summer this way.

At a social function in Brisbane last week I was wearing a light and comfortable Filipino shirt. Many of the men around me were wearing suits and looking red around the gills. I'm not sure how many would be prepared to throw off the shackles of custom, but I can't see that any higher purpose is served by gratuitous suffering in pursuit of European dress habits. Let this be the summer when at last we throw off our jackets and our cultural cringe, shed our ties and our inhibitions and begin to develop our own distinctive Australian dress style. It could combine comfort with ecological virtue, elegance with coolness. Who knows, we might even get Paul Keating and John Hewson out of their imported suits and into something Australian. Then perhaps we could work on doing the same with their economic ideas.

TIME FOR MANAGERS TO SING A NEW TUNE

The Weekend Australian, 25–25 November 1990

There is still a place in modern society for the enthusiastic amateur. Not all that long ago, men's cricket teams in England were divided by a rigid class distinction between the professionals and the amateurs or, as they were known, the Players and the Gentlemen. Those whose inherited wealth allowed them to play cricket simply for fun were segregated from the lower orders who were paid for their efforts. Separate dressing rooms and gates ensured that no perspiring gentleman was contaminated by intimate contact with uncouth sweaty players.

Mind you, in those days our own cricketers were paid so poorly that they might as well have been amateurs. I can remember when Sheffield Shield cricketers were paid less than $2 a day, out of which they had to pay for their own dry-cleaning! Today, the best male cricketers can earn a reasonable living from the game. That seems fair, given the demands of the game at that level and its impact on other career opportunities. However, most of our sports enthusiasts still participate for the love of their chosen game. They are genuine amateurs.

I was reminded of the distinction between amateurs and professionals by a newspaper review of a concert. The reviewer, noting that the Brisbane Chorale had sung a Prokofiev cantata in Russian, praised the dedication of the choristers. But, the review continued, it was unreasonable for unpaid amateurs to go to such trouble. At this point I have to declare an interest, for I was one of those unpaid amateurs. While I have no remit to speak on behalf of other members of the chorale, many of us are there as amateurs in the original sense of that word: we are there because we love what we are doing. We are not paid to do it. Indeed, we pay annual subscriptions as well as incurring travel costs for rehearsals and performances.

Many of our members are professional or semi-professional musicians. Some are conservatorium students who will go on to earn a living as musicians. Knowing my own musical limitations, I regard it as a real privilege to be part of an amateur group which performs with professional musicians. The combination can be very exciting. The arts offer considerable scope for the keen amateur. One reason is that the economics of choral performances are difficult enough at the best of times. Concerts would not be economically viable if the members of a large choir were paid. So as an enthusiastic amateur singer I have been in the ranks of the choir facing famous conductors like Sir Neville Marriner, or standing behind famous soloists like Lisa Gasteen and Luciano Pavarotti. It is also true that amateur performances in music and theatre can have a freshness and enthusiasm which is sometimes missing from professional companies. Again, this is not surprising; many people are more enthusiastic about their hobbies than the work they do five days a week.

On the other hand, there is certainly a class of activities which should be left to the professionals. If I were to need brain surgery, I would want it to be carried out by a qualified professional rather than an enthusiastic amateur. I assume that the pilots employed by our domestic airlines, even after the recent deregulation, will still be professionally competent and not just people who love flying. While the original meaning of the word amateur was one who cultivated an art for pleasure rather than for profit, the word now also refers to a worker who is unskilled or superficial. To call someone an amateur in that sense is certainly not a compliment. For that reason, I was interested in recent criticism of the managers of Australian companies.

A report in *Consuming Interest*, a magazine published by the Australian Consumers' Association, quoted the executive director of the NSW Chamber of Commerce and Industry, Mr David Taylor, on the qualifications of managers. He claimed that only 12 per cent of company managers in Australia hold a university degree and a mere 3 per cent have undertaken formal studies in a management-related field. While the exact figures are a matter of some dispute, there is no argument about the broad conclusion. Much of the managing of Australian companies is being done by amateurs, in the pejorative sense of that word.

Mr Taylor contrasted our situation with the tradition in Japan and the US, suggesting that managers in those countries are much more likely to hold formal qualifications. I recall being startled when I read that shop-floor workers in Japanese car factories are likely to hold Masters degrees in engineering; relatively few managers in manufacturing industry here are that well qualified. This is an important issue because our economic difficulties arise in part from the failure of our managers to handle innovation and the development of quality manufacturing. A recent report by the Senate Standing Committee on Industry, Science and Technology came to exactly this conclusion: our managers tend to be inflexible, short-sighted and threatened by change.

Consuming Interest claimed that our managers lag behind big foreign competitors in terms of education, innovation and flexibility. Inexperience in strategic planning and human resource practices were noted particularly. It also concluded that our managers are 'definitely not setting world standards for social and environmental responsibility'. Some recent spectacular corporate failures have posed obvious questions about management standards. While there is a role for the keen amateur in the performing arts, if we want our industry to perform we need a professional approach from its managers.

THE SEVEN AGES OF CRICKET

A personal reflection, 2009

All the world's a stage,
And all the men and women merely players;
They have their exits and their entrances;
And one man in his time plays many parts,
His acts being seven ages. At first the infant,
Mewling and puking in the nurse's arms . . .
    Last scene of all,
That ends this strange eventful history,
Is second childishness and mere oblivion;
Sans teeth, sans eyes, sans taste, sans everything.
(William Shakespeare, *As You Like It*, Act 2, Scene 7)

As usual, Shakespeare said it all perfectly several hundred years ago. I find myself reflecting on his eternal wisdom as my cricket career comes full circle. I have been what is now called a 'cricket tragic' for most of my life. Unlike at least one professed cricket tragic, I can actually play the game as well as talk about it. It all started more than 50 years ago. Huddled under blankets late at night, listening to the crackling radio coverage provided by the ABC, I heard the events unfold at Old Trafford in 1956 (when Jim Laker skittled Ian Johnson's team, becoming the first man ever to take all ten wickets in an innings after having taken nine of the ten in the first innings, an extraordinary feat by any standards). As a weedy bespectacled youth growing up in the New South Wales bush, I went to the scruffy town ground whenever the local team was playing, wearing my white shorts and sandshoes; I knew that my offer to keep score would be gratefully accepted but I always hoped somebody would be late so I would be asked to field. The first time this happened was every boy's dream; as the skinny young substitute I was sent to fine leg, with instructions to try to stop anything hit in that direction. In the first over the opposition opening bat miscued an attempted pull shot and I sprinted in to catch the ball just above the ground. The batsman concerned was a local legend who regularly scored centuries against our long-suffering bowlers and went on a few years later to open the batting for NSW Combined Country against the finest Sydney could put in the field, so I was briefly a hero. Five minutes later, the missing player arrived and I was banished back to the scorer's table. But I bowled at home for hours at the front gate and threw myself catches from the back steps, hoping to become good enough to play with the big kids one day.

By the time I was 16, I was playing occasionally for the local team as a determined but strokeless bat and tidy off-spin bowler. My first significant achievement was to stonewall for two hours to help salvage a draw after all the real batsmen had been dismissed. In those days, the team for 'away' games was chosen by picking three or four people who had cars, then filling in around them. As our captain (and opening bowler) weighed 23 stone (about 150 kilograms!), this had to be taken into account in loading the cars. I was chosen more often for away games as my light weight made me the ideal person to sit on the same side of the car as our very large bowler. Then I had my growth spurt and discovered I could bowl quite fast with useful outswing, as well as having safe hands and reasonable reflexes. This made me an automatic choice for the local team and we had two quite good seasons.

Selection as an opening bowler for the District Under-18s followed, then my first premiership as an 18-year-old, following up my three wickets with a tense one-hour ninth-wicket partnership of 34 to see us home. We were promoted to A Grade and I had a very good season with the ball. I was called up to the District XI when the usual opening bowler broke down and made a striking debut; 3-11 from an opening eight-over spell before I tore the heel off my left boot ensured my retention in a team that went on to be undefeated interdistrict champions. By then I was working in Sydney and studying part-time, so I started playing for the University of NSW. In six seasons there I was in three premiership teams. My last contribution to the club, seven wickets to help win the 1968 final, left me agonisingly on 199 wickets for UNSW in grade matches; the captain dropped a steepling miscue at mid-off and tried to redress this by giving me a long spell in the second innings but the elusive 200th wicket would not come. I also played intervarsity cricket against many Queenslanders who went on to play Sheffield Shield (Bill Buckle, Lew Cooper, Bob Crane, Sandy Morgan, John McLean, John Loxton) as well as future Test players like John Inverarity, Tony Mann and Rod Marsh, leading to my selection in 1968 for Combined Australian Universities. Three years studying for a doctorate at University of York also involved intervarsity cricket against future Test players like Frank Hayes, selection for the British Universities squad, Yorkshire League and local cup games. I had a fair bit of success as a bowler of very lively pace who genuinely swung the ball. Highlights were 7-12 to rout Lancaster University for 27 in the annual Roses match, and 6-12 to clean up Dringhouses in the final of the Yorkshire Senior Charity Shield. Moving south to work for the UK Open University, I played eight seasons of serious club cricket as well as occasional appearances for a travelling side called the Incogniti, an association that has continued with me joining them more recently on tours to Western Australia, the Indian province of Goa and in 2003 the Channel Islands. The highlights of this period included playing on the county ground at Taunton against Somerset Stragglers and the only time in my cricketing life that I took all ten wickets. It was a classic reminder of the arbitrary way fate allocates its favours. Before I took the first wicket in that match, for the Open University against the local village of Aspley Guise, our wicketkeeper dropped a routine catch from the other opening bowler. So you have to be very lucky to take all ten wickets. Yes, you have to bowl well – but you also need the incredible good fortune of having nobody dismissed in any other way during the entire innings.

When I moved to Brisbane in 1980, I found some of my old intervarsity foes were still playing club cricket for University of Queensland, so I played two seasons for them, taking nearly 100 wickets and playing in a team that won a lower-grade premiership. I started playing for Griffith University in the Warehouse competition in 1982–83 and still turn out when I catch the selectors' eye. Over 27 years with Griffith, I have taken quite a few wickets and played in several premiership teams, as well as scoring my first (and last) century as a promising youth of 42. I had several good days on which I took six or seven wickets and one where I took eight. I was annoyed on that occasion, however. I had taken eight wickets on two earlier occasions in my life, and once took all ten, as mentioned before. I had hoped I might take nine one day to complete the set! I had taken eight of the first nine that day and had the prospect of four deliveries at the last man – but his captain had so little faith in number 11's ability to survive that he declared their innings closed without allowing the batsman to come in.

In recent years I have also played a bit of veteran's cricket, where the bowling is a bit more friendly for an old bloke; I have turned out for University of Queensland Old Boys in local fixtures and the first two Vintage Cricket carnivals, in Adelaide and Perth, while I have also played with Sunshine Coast Antiquarians in local games, two Golden Oldies carnivals and the 2006 Barbados Vintage Cricket carnival where I had the privilege of opening the bowling on my 64th birthday (and taking three wickets!). I have also toured the Christchurch region four times with a wonderful group of senior cricketers called the Honest Trundlers, who now return there every February. And in 2009 I had the honour of playing for Queensland in the national over-60s carnival in Tasmania. As one of my fellow geriatric cricketers put it, it's just like the game we used to play except it's in slow motion! The bowling is slower, the ball is not hit as hard, the fielders move more slowly.

These days I am rarely in the best XI that my Griffith mates can put on the field, but (like in the 1950s) I still always take my gear in the hope that somebody won't show up and I will get a run. As the Bard put it, 'Last scene of all, that ends this strange eventful history, is second childishness and mere oblivion.' In my 50th season of adult cricket, the team let me open the bowling when my 65th birthday coincided with a match day; unfortunately I bowled badly and presented easy runs to the lucky batsmen! But the next day, in a charity match at Buderim, I showed there was still some life in the old dog by batting for an hour in a partnership that added 120 and made our score respectable, then opening the bowling with a tidy spell. So, second childishness perhaps, but not yet oblivion. As a few of my Griffith mates wanted to end their cricketing career on a high note, I sent emails to cricketers I had played with and against around the south of England and put together a great tour for August 2008, seven games in 12 days on beautiful grounds. It doesn't get much better than that; touring with mates who have shared the highs and lows of this wonderful game for 25 years. The great thing about cricket is that it is essentially a team game. While it allows for individual achievement with the bat, with the ball or in the field, everything you do is based on the support of your mates. And your role changes over time, but there is still a niche. I was inspired to find myself playing in a Golden Oldies carnival against a Kiwi who gave up bowling at 70 but was still taking the field at 82. He later filled in for us when one of our players broke down in the Golden Oldies carnival at Queenstown, by then a lively 85-year-old. There's an 'aspirational target'! It has not escaped my notice that my 70th birthday will be a Saturday in the 2012–13 season . . .

When I was growing up in the bush, almost all adult males (and most of the females) played sport. The younger men played tennis or cricket in summer and some form of football in winter. Those of more mature years mostly played golf, while those even older usually took up bowls. One obvious reason was that being a member of a sporting club was really the only legal way to have a beer on Sunday afternoon! But it meant sport was the social glue that held the community together. Participation in sport has gradually declined over my lifetime, making us a less healthy community both physically and socially. Other changes, like the increasing tendency for businesses to open at weekends, have reduced the fraction of the adult population able to play sport. So we are steadily becoming less active, less fit and more likely to be overweight. That is not a good trend. All of the indicators show that there is a positive effect on health outcomes of remaining physically active, just as there is a positive effect of remaining mentally active and learning new skills. It would be a good investment in our future, almost certainly highly cost-effective in terms of savings on the illness budget for hospitals, to develop systematic programs to encourage those of mature years to stay physically and mentally active.

EDUCATION

INTRODUCTION: EDUCATION AS THE KEY TO OUR FUTURE

Knowledge will forever govern ignorance. The recent economic success stories in our region are countries like Singapore that have few natural resources but have invested in the resource of their people. As was discussed in the section on economics and politics, the modern economy is dominated by what Barry Jones called 'brain-based industries'. So we should see education not as a cost but as an investment, arguably the best possible investment in our future. The education and skill levels of our community will determine our capacity to respond to new challenges.

As discussed in the section on culture and health, community behaviour can be altered by legal force, by financial incentives or by education. I believe education is the most important of these factors. If our attitudes have changed, we will behave differently without being coerced by legal force or enticed by economic incentives. If our attitudes have not changed, we are likely to resist what we see as unjust laws and ignore what we see as irrelevant incentives. So the direction of 21st-century Australia will be significantly influenced by education. This naturally includes the formal education of schools, technical colleges and universities. But it also includes what could be called informal education.

Let me give you a specific example. I argued in the 1991 Boyer lectures that television has played a major role in educating our young people. From television they have learned that our society values the collection of material goods above any other measure of success. They have learned that a simplistic assertion is more impressive than an honest attempt to wrestle with the complexity of the real world. And they have learned that physical violence is a perfectly acceptable way, perhaps the best way, to resolve differences of opinion. There are, of course, also benefits of television. We have the opportunity to be much better informed about the world than any previous generation. The best of television is a powerful tool for real, positive learning. As a concrete example, both of my sons reached primary school knowing more biology than I learned in all of my formal education, through watching the wonderful films made by David Attenborough.

I have argued in this collection that we face new challenges both more complex and more demanding than those faced by any previous generation. While the formal education system has to respond, we should also be aware that most Australians are beyond the reach of that system. So we need to develop our informal education to ensure that all of the community is equipped to play a positive role in shaping our future. The writings I have collected in this section consider various aspects of the informal and formal education we need to fashion a sustainable future.

MY PERSONAL JOURNEY FROM CARAGABAL TO COPENHAGEN

At Caragabal primary school, I was already aware that being what we would now call a smart-arse was not a good way to be popular with your classmates. Being younger and smaller than your classmates meant that being quicker than them to come up with the required answer was not a good move. Retribution was usually swift and physical.

When I was eight, my parents moved to another small town much closer to Sydney. When the new school year started and I enrolled at Tahmoor primary school, also a two-room institution, I was shocked to find that my new classmates in Grade Four had done things I hadn't. So instead of being ahead in arithmetic and social studies, I was some distance behind. I soon caught up and by the end of the year was ahead of the class. In Grade Five, I progressed so rapidly that the headmaster suggested to my parents that I should sit the public examination for a state bursary. This involved considerable extra reading and practising arithmetic calculations, but I relished the challenge. Allowing a child to sit for the bursary examination was not lightly undertaken, as the upper school classroom had to be dedicated to that purpose for a whole day, with all the other pupils in Grades Four to Six given a day off school. On the appointed day, I was seated alone in the middle of the room under the constant gaze of three local senior citizens. The printed examination papers had been sent from Sydney in large envelopes with impressive seals. For each of the three written papers and the dictation test, the seal was broken with great ceremony. I was allowed short breaks between the tests. My written answers were placed in another large envelope, then also solemnly sealed with red wax and sent off to Sydney for marking. I was physically sick the next day from the stress of the whole process. But great was my joy when the letter from the department to my parents told them I had been awarded a bursary to help with the expenses of my secondary schooling. There were conditions; the most significant was that I had to study a foreign language. This in turn meant that I could not attend the local central school, which had three years of secondary classes, but would have to attend the big high school 25 miles away at Bowral. I believe there were long discussions between the primary school headmaster, my parents and the district inspector about the wisdom of allowing a child barely 11 to start high school, especially as that would mean leaving home each morning at 6.30 am to catch the school bus and not getting back home until 5.15 pm. I gather they were reluctant to approve that sort of transition and wanted me to spend another year at primary school. Having already marked time for a year in Grade Three, I was horrified by the prospect of spending another year doing relatively little, compared with the tantalising prospect of going to high school and embarking on the study of exotic foreign languages and new areas like science. Against their better judgement, they let me start high school.

There is no doubt I was intellectually ready for high school, as I soaked up the exotic new subjects like blotting paper, but I was quite immature and unprepared for the rough-and-tumble of a big rural high school. As an example of my naivety, when I was summoned out of a maths lesson, escorted to the room where 3B were being taught English and asked to identify the clause underlined in a sentence on the blackboard, I innocently said it was an adverbial clause of reason and was sent back to my maths class by the smiling teacher. As I extricated myself painfully from a large holly bush at lunchtime, I had cause to reflect on the wisdom of my action, which had allowed the English teacher to berate pupils three or four years older than me (and much bigger and stronger as well) for not knowing the sorts of things that little first year kids knew. As I was the youngest, smallest and least physically developed student in my year, I was probably destined to have a rough ride. That became a certainty when my first year maths teacher detected that I was seriously short-sighted, as a result of which I returned after the Christmas holidays wearing glasses. Not just skinny and weedy but also wearing glasses, I was destined to be at the bottom of the social heap for the entire five years of secondary school. The compulsive overachieving workaholic of my adult years was probably shaped by those formative years.

As one of five children of a rural family, there was no way my parents could fund me to go to university. I managed it by finding a position as a cadet electrical engineer, working in a factory which produced equipment for the telecommunications industry. I was given one afternoon a week off work to attend the NSW University of Technology, mainly based at the Sydney Technical College in Ultimo but slowly building on its new campus in the sandhills of the old Kensington racecourse. After nearly four years of that work, I successfully applied for a job to use my electronics expertise building equipment for a research group in the physics department of that university, by then expanded and renamed the University of NSW. I soon recognised that the physics was more interesting than the electronics, so I transferred to a science degree. As part of the humanities program which was then compulsory, I also studied English language and literature, history and politics. I did well enough in final year physics subjects to qualify for the honours program, an additional full-time year, which I funded by teaching first-year students for 12 hours a week. I think, in retrospect, that I learned more physics teaching it to undergraduates than I did bluffing my way through exams!

After graduating, I worked for three months in the Defence Standards Laboratory, went to the US for ten weeks on a travel program funded by the Department of State, then taught high school science for a year and a half, again widening my understanding of science by being forced to teach subjects I had never studied, biology and geology. I was offered a scholarship by the University of York to pursue research for a doctorate; given where I am now, it is a touch ironic that the work was funded by the UK Atomic Energy Authority!

In the process of writing my thesis, I was offered a one-year temporary lecturing post at the new UK Open University. The new institution taught mature-age students wherever they were in the UK by preparing printed materials, TV and radio programs, residential summer schools and local tutorials. As all courses were prepared by multidisciplinary teams, working there was a wonderful learning experience. The courses in technology went well beyond the traditional technical content to include economics, political issues, environmental questions and the broader social process of technological change. It was wonderful to be teaching adults with lots of life experience, using materials that reflected the complexity of the real world.

Five years into the appointment, I was offered a fellowship to spend six months teaching the field of science, technology and society at Griffith University, then a very new institution with a wonderful bushland campus in the southern suburbs of Brisbane. The very first intake of students was still completing the undergraduate degree. The university had decided that all science students would study the history and philosophy of science as well as the social structure of the scientific community and the role of science in economic development. It was part of the whole university's multidisciplinary ethos, recognising that all the complex problems of the real world defy being compressed into the boundaries of traditional academic disciplines. That was clearly an innovative move ahead of its time. Thirty years later, no other Australian university requires its science students to study this broader context for their work, while the Griffith commitment has been steadily watered down by academics seeking a return to earlier, less complicated days.

I went back to the Open University in the middle of 1977, but returned to Griffith at the start of 1980 to take up a permanent appointment as a senior lecturer and director of the science policy research centre. I have stayed at Griffith University, where I was promoted to associate professor and then professor. When I negotiated voluntary early retirement at the end of 1999, the university made me an emeritus professor, a lifetime honorary appointment. I continue to give lectures at the university's Nathan campus, now only one part of a multi-campus institution. I have adjunct appointments at other Australian universities and every year give guest lectures at several institutions. Although I am now past the traditional age of retirement from full-time work, I am still actively learning and trying to pass on my new insights. In the complex world of today, where knowledge is continually expanding, we all have to keep learning.

Many of the tasks I have undertaken outside the university system have required rapid learning. When I was appointed to the National Energy Research, Development and Demonstration Council, I knew a fair bit about social and environmental issues related to energy but very little about the production of coal, oil and gas. When I was asked to direct the Commission for the Future, I had a strong background in the technologies and environmental issues but had to learn very rapidly about social issues like the ageing population. When I was invited to chair the advisory council that produced the first national report on the state of the environment, I was on a very steep learning curve in some of the ecological areas. Every piece of learning equips us to be a better learner when faced with the next challenge. A learning society – the only positive future – needs to be based on us all being individual learners, committed to maximising our own potential and committed to participating in that society.

LET'S EQUIP THE YOUNG FOR THIS WORLD OF RAPID CHANGE

The Weekend Australian, 23–24 June 1990

All should welcome serious debate about the most appropriate education system for the future. Education is the key to the lives of our young people, as well as to our future economic prospects in a world of rapid change.

I am sceptical of general assertions that standards are falling. Such claims are often based on nostalgic half-remembered impressions of an earlier golden age of supposed rigour, discipline and valiant endeavour. As an example, Tim Dare's article (in today's paper) suggests that the NSW system of 30–40 years ago was 'job-oriented' and so 'employers got what they wanted'. That doesn't square with my memory of a diet of Latin irregular verbs, Euclidean geometry and obscure English monarchs. In those days of full employment, there were jobs even for those who left school with no discernible skills, little useful knowledge and inadequate personal hygiene. You have to feel sympathy for 15-year-old school leavers who can't get a job because they can't spell, given that most 50-year-olds who can't spell have always had jobs.

While we should be concerned that some young people leave school inadequately equipped to face the world, we should not be deluded into thinking that this is a new problem. The education of the 1950s was strongly oriented toward external examinations. The NSW selective schools had talented students and good teachers. If the results in the external examinations indicated educational achievement, then most high achievers came from the selective high schools. We should be cautious about attributing better results to the schools, however, as it has been shown that most such differences are explained by the students' socio-economic backgrounds.

The debate is not helped by sloppy use of statistics. Tim Dare correctly points out that it is not possible to make a simple comparison between government and private schools, as the latter set consists of two distinct groups: the Catholic system of schools and the 'independent' schools. This caution is exemplified by Dr Partington's dubious conclusion (in this paper recently) that parents are moving their children to private schools despite what he claims to be better facilities and higher teacher–pupil ratios in the government schools. As a group, Catholic schools are poorly equipped and staffed by comparison with government schools, but many of the independent schools are lavish. There is no obvious basis for a sweeping conclusion that government schools are more generously funded than private schools. It depends which private schools are being used for the comparison.

A case can be made for the view that schools are failing our young people, but this does not support the proposition that we should go back to previous education practices. There is little to be said for NSW Minister for Education Dr Metherell's proposed great leap backwards, with its talk of 'tough core subjects' such as maths and languages. I went to school in the bygone days when beaches were clean and books were dirty. We were told that only very clever children could learn

French. Nobody thought to ask why even stupid French children are able to learn their language! There is nothing intrinsically difficult about learning languages, although the way they were taught then usually made them seem hard. What the Metherell approach really does is select a small elite by making learning unnecessarily difficult for most young people.

A more appropriate aim for education is to equip all young people for a world of rapid change. We need to move away from our traditional obsession with the content of the curriculum, based on the notion that there is a fixed body of facts which able students should learn and a defined set of skills for them to acquire. Those who are currently starting school are likely to be working until about 2050. Their schooling cannot possibly give them the skills and knowledge they will need between the years 2000 and 2050, since much of that knowledge and many of those skills do not yet exist. Many of the 'facts' we might teach them now will eventually be seen to reflect our current primitive understanding of the world.

We need to recognise that schooling is not the last word in education; it must be the first stage in a process of lifelong learning, whether that includes tertiary study or not. So the process of schooling needs more attention than the content. We should move right away from what has been called the Moses model of education, the delivery of tablets of stone to be memorised. We are much better served by a cooperative learning model, in which teachers and their students grapple with complex questions for which there is no correct answer at the back of the book, or anywhere else for that matter. There is no 'right answer' to such questions as how we should handle plastic waste, or what we should do to slow down the rate of climate change, or how we can really help poor countries in Asia to raise their material living standards.

The problems of the modern world do not fall neatly into the divisions of the traditional disciplines. In the real world, science is mixed up with economics, geography is mixed up with history, and maths is used to obtain partial solutions of messy problems rather than neat answers which are always round numbers. This does not justify a woolly populist approach in which any half-baked answer is equally good. Producing an acceptable answer to a complex problem is a more demanding task than trotting out a well-rehearsed proof for an obscure theorem.

The reason that the new approach works is that most students are motivated by complex real problems in a way that they were not by obscure academic abstractions. The old system served those of an academic bent and those for whom education was a game, a set of hurdles to jump for the reward of being allowed to attempt even higher hurdles next year. For the majority, a problem-oriented approach provides a more secure foundation for the world they will face: a world of rapid and unpredictable change. Even occupations which were once boring and repetitive are now increasingly likely to demand cooperative problem-solving skills. The Victorian approach contains some of the features needed in an education for the future; the Metherell scheme, alas, is distinctly an education for the past.

EDUCATION FOR SUSTAINABLE FUTURES

Prepared in 2005 for the Queensland Ministerial Council on Educational Renewal

About ten years ago, I visited the Shinto shrine at Ise in Japan. This large and immaculate wooden building is the main temple for that religion, the equivalent of St Peter's in Rome. There are two identical sites, side by side, in a forest that produces enough wood to allow rebuilding every 20 years. So the shrine is constantly renewed. Every 20 years a replica is built on the vacant adjoining site, the sacred objects are transferred and the old building pulled down. The process of constant renewal, in balance with the capacity of the local ecosystem to supply materials, has continued for more than 1200 years. It is a wonderful metaphor for a sustainable society. As long as we remain in balance with natural systems, we can renew and upgrade our structures indefinitely.

Sustainable futures will involve significant change. The 1987 Brundtland report said the world's economic and environmental futures are intertwined and should be seen as complementary, rather than competing. In 1992 COAG adopted a National Strategy for Ecologically Sustainable Development, defining sustainable development as meeting the needs of Australians today 'while conserving our ecosystems for the benefit of future generations'. In those terms, sustainable futures are simply common sense: enjoying the fruit without damaging the tree.

In operational terms, a sustainable society would not be eroding its resource base, causing serious environmental damage or producing unacceptable social problems. Our present lifestyle does not satisfy *any* of those main criteria. We are wasting resources future generations will need, especially oil, damaging environmental systems and reducing social stability by widening the gap between rich and poor, both within our country and globally. We now know that inequality does not just produce social stress, as Wilkinson has shown that it has measurable ill-health outcomes. If civilisation is to survive, the next century will have to be a time of transformation of our technological capacity as well as our approach to the natural world, and to each other. Jared Diamond argues that societies expand until they reach limits, then survive or fail depending on their capacity to adapt to the new challenge. History shows that some collapsed, like Easter Island or the Mayan civilisation, while others adapted and survived, like other Pacific islands and medieval Japan. Diamond shows that our choices will determine whether we can achieve sustainability by getting back into balance with our resource base. The alternative is dire.

A sustainable future has to have an assured resource base, respect the limits of natural systems, be socially stable and economically secure. A critical debate concerns the relationship between the economy and the environment. Herman Daly, the pioneer of 'steady-state economics', argues that it is impossible for growth to continue without limit in a closed system and urges us to move away from seeing the scale of growth as the defining characteristic of economic health. On the other hand, the World Business Council for Sustainable Development argues that sustainable development is good for business and business is good for sustainable development. Some economists think it is possible to merge the concepts of sustainable development and competitiveness, so there does not have to be a trade-off. There are many opportunities for improved design to reduce the resources needed for the goods and services we use; using resources efficiently could increase competitiveness and so enhance economic prospects. However, unless we recognise that perpetual growth is not possible in a closed system, improving efficiency will only defer our inevitable collision with the limits of the natural world.

The second report in the UN series on the global environmental outlook, *GEO2000*, noted that the present course is unsustainable, so doing nothing is no longer an option. We face serious challenges: loss of biodiversity, the state of many waterways, loss of wetlands, degradation of productive land and the risk of catastrophic climate change. The third report explored four possible futures for the world. In Markets First, effectively the present approach, globalisation and a liberal trade agenda promote rapid economic growth, but nations are increasingly unable to prevent worsening environmental damage and growing political instability undermines the conditions for orderly economic development. In Security First, the wealthy use force to try to suppress growing concern about ecological problems and a widening gap between rich and poor, creating a divided and violent world. In Policy First, governments take decisive action to curb environmental excesses but this doesn't raise material living standards of poor countries to an acceptable level. The most hopeful scenario, Sustainability First, is based on a shift in *values* and setting a goal of satisfying basic needs for all within the limits of natural systems. There are two obstacles: the present global society is a long way from having those values and we don't yet know how to interact sustainably with natural systems. Great changes can in principle be made by policy reform, which could dramatically cut the resource demands and environmental consequences of our lifestyle, but the political will to implement such a strategy is not in sight. Until politicians appreciate the urgency of our predicament, there won't be the political will to embrace a different approach.

The Global Scenarios Group elaborated on this theme in their influential report *The Great Transition*, arguing that the move to a sustainable society involves new values. As discussed earlier (see page 91), Paul Raskin argued that the values underlying recent development are domination of nature, consumerism and individualism. These values, he said, are incompatible with the goal of a sustainable future. Domination of nature should be replaced by a willingness to live within the sustainable limits of natural systems, consumerism by an emphasis on quality of life, and individualism by a renewed sense of community, recognising that we share a common fate with the whole human family. These principles would be applied with different weights and shades of meaning in different societies, but they must be the underpinning values. In this plausible future, the economy is seen as a means to social and cultural ends, rather than an end in itself. The overall scale of the economy is much greater than it is now, so people would be better off financially, but the flow of material resources is far less, water is used sustainably and fossil fuel use has fallen dramatically. This imagined future is much more equal than today and participatory democracy exists at the local, regional and global level. This is an inspiring vision of the sort of world we should be aspiring to build.

There are three ways to change human behaviour: economic inducements or penalties, law or regulation, and education. Since resource use is relatively insensitive to price, and legal restraints are only effective if they are consistent with community values, I believe that education is the key to becoming a sustainable society. Peter Sterling argued that education must play a central role in the unprecedented changes that we need and education will itself be transformed in the process. Education does not just make people aware of the problems; it should also give them the capacity to find solutions. So education for sustainability must include, in terms of content:

- understanding of the impacts of human activities on natural systems;
- awareness of the finite scale of non-renewable resources;
- awareness of the limits on use of renewable resources;

- understanding of the need for durable economic activities;
- awareness of the consequences of increasing social inequality; and
- understanding of the importance of cultural traditions, beliefs and practices.

This is the cognitive basis of education for sustainability, but it is not the whole story. In terms of developing capacities, education should aim to give young people the ability and confidence to shape their own future, rather than passively accepting a future imposed on them by forces they cannot control. It should also aim to instil an appreciation of our moral responsibility toward other species and future generations.

Perhaps the best metaphor for the outcome we should be seeking is *Globo sapiens*, as introduced in 'Social and environmental impact assessment: A case study of the hydrogen economy' (page 68). As Patricia Kelly writes, like our students, we struggle with our values. We face the challenge to transform ourselves into teachers who are also wise global citizens, aware of our responsibility to humanity, to the other species that share this planet with us and the future generations for whom we hold it in trust, if we want to develop these attitudes in our students. So we need to be on a learning journey to develop the characteristics of *Globo sapiens*: empathetic, globally conscious, thinking beyond our generation, courageous, willing to embrace the changes involved in working for healthier futures. We must also ensure that our schools don't just teach the cognitive elements of sustainability, but embody its values. Schools should be models of best practice in their buildings, use of energy, water and resources, in the way their grounds value and protect biodiversity, and in their social inclusiveness. Then we will really be educating for sustainable futures.

THE ROLE OF UNIVERSITIES IN THE MODERN WORLD

Edited extract from Our Universities are Turning Us into the Ignorant Country, *NSW University Press, 1994*

Universities fulfil a range of functions. They conserve knowledge through libraries and scholarly collections. They transmit knowledge to students, in many cases by announcing theories or facts for the students to memorise, in others through guided learning programs and by allowing students access to their libraries and scholarly collections. Universities advance knowledge by basic research, once the hallmark of a university but now seen by government and some of our older institutions as an optional extra for some of the academics in a minority of universities. They apply knowledge by consulting and applied research; these activities are almost universally prized because they bring extra resources into the system. Although it is now being seen as an indulgence of questionable importance, universities also refine knowledge through critical review and scholarship. Accreditation of professional qualifications in fields such as medicine and engineering is largely the responsibility of universities, which attest to the public and the profession that graduates have at least achieved minimum standards. Finally, university academics have a duty to act as the conscience and critic of society. In the dark days of the Bjelke-Petersen regime in Queensland, it was only a handful of university staff and a few dissident clerics who stood up against the flagrant abuses of state power and the blatant corruption in high places.

All the functions listed above are important aspects of the work of the university. All should be fostered, given resources and rewarded. Staff who are especially good at any of these functions should be considered for promotion and more generous pay. The time and energy of staff are limited resources. While staff can be made more or less productive by changes to their working environment, extra effort in one area can usually be achieved only at the expense of others. Pressure to give attention to any one aspect of the role of universities, in the absence of extra resources, must inevitably mean that the other contributions of the university system to society are eroded.

All universities are nominally committed to excellence, but it is hardly a distinguishing characteristic; airlines and even fast-food chains claim to seek excellence. I am concerned by the impacts of recent policy changes on the university system as a whole. The increased emphasis on undergraduate teaching and applied research has caused a significant erosion of several of the other functions of universities. We have systematically reduced the capacity of universities to conserve knowledge (through libraries and scholarly collections), to advance knowledge (through basic research), to refine knowledge (through critical review and scholarship) and to act as the conscience and critic of society. This represents a considerable loss. These functions are at least as important as producing more graduates or increasing commercial exploitation of new knowledge. In the modern university, they are typically neither valued nor resourced.

The expansion of numbers in higher education without a corresponding increase in resources in general and academic staff in particular has seriously eroded the quality of the educational experience offered to our students. This is obvious to all who work in higher education. While I am particularly aware of the changes forced on science teaching, I get similar reports from a wide range of disciplines: less small-group teaching; less individual attention; less practical experience; fewer opportunities for students.

The changes of the last decade to the pattern of funding for universities have produced a significant incentive to lower standards. There are now enough examples in the public domain for it to be clear that incentive has been accepted by some parts of the system, seriously devaluing the currency of Australian degree qualifications. We will all suffer when the extent of this new 'client focus' becomes apparent to those in other countries who have traditionally respected the standing of our graduates.

The complex problems our society faces in the 21st century require professionals with the capacity to work across traditional disciplines in problem-oriented teams. The changes of the last decade have seriously eroded the capability of universities to produce such graduates. The rapid pace of globalisation demands that we produce graduates with globally-portable skills and the capacity to respond swiftly to a changing world. Again, under-resourced universities cannot respond adequately to this challenge.

All the overseas evidence confirms the view that investing in a well-educated workforce is the best possible way of assuring our nation's economic future. The HECS scheme gives the impression that university education should be seen by individuals as a personal investment in their own future earning capacity. This is driving student choice away from the capacities that are needed in our broad national interest toward the skills that students judge will maximise the probability of achieving a financial return on their investment in fees. So there has been a drift away from such fields as science and engineering, which are seen as being intellectually demanding without the carrot of future financial rewards, towards law, medicine and business studies. Making HECS repayment a future taxation liability constitutes a real, tangible incentive to our brightest graduates to seek overseas employment in preference to working here. The incentive is clearly working as increasing numbers of graduates see the financial benefit of staying overseas.

The current state of the university system represents the result of a long series of ad hoc decisions driven by short-term political expediency. The overall result is that we have seriously eroded the capacity of universities to provide the sort of graduates we need for the coming century. It will not be easy to remedy the problems, but we have a chance of making progress if we recognise them and begin the task of investing in our common social and economic future.

THE NEED FOR ENVIRONMENTAL LITERACY

Edited form of 2004 conference paper

Universal literacy has been an educational goal for many decades, so Australians now live in a society in which most people can read and write. There are, however, still literacy problems associated with the disappearance of grammar from the formal curriculum. Many young people whose oral communication skills are acceptable have such a shaky grip on grammar that their writing barely meets minimum standards. While poor literacy was not an impediment to finding gainful employment 50 years ago, it is now a serious obstacle because technology has gradually removed many of the jobs that did not require any form of written communication. We are now in the last stage of the gradual transformation of human society, from the early days of literacy when a few savants who could read and write became the custodians of knowledge, through successive stages of expansion of educational opportunities which have widened the group of people who are functionally literate. Modern society assumes that we can all read and write; those who cannot use the local language fluently, such as recent migrants from other language backgrounds or those whose education was severely deficient, are at a huge disadvantage.

I argued a decade ago that scientific and technological literacy were also required in a modern society, where our lives are affected so much by developments in science and technology. The argument for environmental literacy is similar. The growth of the human population and the increasing power of our technology mean that we are no longer just one of several million species inhabiting this planet. We are now an active agent of physical, chemical, biological and geological change. Our burning of fossil fuels has changed the capacity of the atmosphere to trap heat and so changed the climate. The clearing of vegetation and the covering of huge areas with tar and concrete have changed the amount of the Sun's heat absorbed by the Earth. The production of huge amounts of chemicals has changed the chemical balance of the air, the oceans and the soils; the IGBP Report said that the human interference in the nitrogen cycle, mainly by taking the gas from the atmosphere to make fertilisers, will be seen in future to be as serious as our disturbance of the carbon cycle, which is changing the global climate. The driving of some species to extinction, the release of exotic species and recently the production of new genetic identities have all changed the biological balance of natural systems. Finally, we have changed the course of rivers, built reservoirs and artificial harbours, influenced sand movement along coastlines and in other ways changed our geological surroundings.

As the first Australian report on the state of the environment said, our serious environmental problems are the consequence of the scale and distribution of the human population, lifestyle choices, technologies used and the consequent demands on natural systems. In other words, everyone now makes decisions that have implications for the natural system – as a worker, as a consumer, as a parent or as a member of a community group. Our urban structures, our legal system, our economic development choices, our use of transport, our recreations and amusements, our diet and the way we live our daily lives all have significant impacts on the natural environment. The argument for universal environmental literacy is simply an argument that we should understand the effects of our choices, rather than continuing to do unnecessary damage through our ignorance.

Recognising the importance of aiming at widespread or universal environmental literacy has implications for both the content and the process of education. In terms of content, we need to aim at an understanding of the underlying science of our interaction with natural systems, but that understanding needs to include the complexity of the questions and the consequent limitations on our knowledge. In other words, the limits of our present knowledge mean that scientific knowledge could be described as islands of understanding in oceans of ignorance. Science is, in the terms of that metaphor, always engaged in land reclamation, but there is no prospect of filling in the oceans of ignorance in our lifetime. Pursuing that metaphor one step further, an enduring problem is that islands of scientific understanding have been seen as separate entities which are not connected. So agronomists have expert knowledge of pastures, but may not understand the implications for surrounding bushland of changes to the pattern of land use on farms. Foresters have detailed knowledge of the managing of wooded land, but may not know about the effects on river systems of changes to the way we use our forests. Transport experts may be able to build roads or even design overall systems of urban transport, but may not understand the effects of the resulting travel on the social dynamics of the city, on local air quality or on the global climate.

Recent international efforts have been aimed at sustainability science – a style of scientific inquiry which explicitly recognises the complexity of natural systems and the resulting need for interdisciplinary study to improve our understanding of those systems, as well as taking into account the complexity of human interaction with those systems. Thus, for example, the way farmers use the land is being affected by the changing global climate, but one of the factors changing the climate is the way farmers use their land, so our developing understanding of the problem needs to take into account the links in both directions between the local and the global. The key *content* of education for environmental literacy is probably what Barry Commoner called the Four Laws of Ecology: everything has to go somewhere; everything is connected to everything else; there is no such thing as a free lunch; and nature knows best. Most of our serious environmental problems arise directly from a failure to understand those basic ideas. While there is also value in making all Australians aware of, for example, our unique biological diversity, traditional education has often concentrated on the individual trees rather than the nature of the wood. The overarching principles are much more important than the details of particular species or habitats.

That being said, we still urgently need a better understanding of the local biota. It has been estimated that we have only identified about ten to 15 per cent of the million or so species found in Australia. Even at higher levels of organisation such as vascular plants and vertebrates, we are still encountering species that were previously unknown such as the Wollemi pine, a tree growing to 35 metres in height within 100 kilometres of Sydney. As taxonomy is not seen either as an exciting area of science or as a high priority for research resources, the rate of progress is alarmingly slow. It is estimated that it will take hundreds of years to identify all the plant and animal species of the continent if we continue to proceed at current rates. There is no prospect, even in principle, of understanding the impacts of our actions on those species we have not yet even identified.

While 85–90 per cent of the species living here are unknown, many of the others are not well understood; they have simply been identified and described in enough detail to allow recognition. Again, there is no realistic prospect of understanding all the impacts of our actions on the species whose characteristics and behaviour remain largely a mystery. Without an improved understanding of the basic building blocks of the natural systems of Australia, we cannot hope to interact sustainably with those systems.

We also urgently need a better understanding of complex systems. It is now clear that many of today's environmental problems stem from past well-intentioned advice, whether to irrigate arable land or to clear vegetation or to introduce exotic species. While each research project extends our knowledge base or clarifies our understanding of some parts of the system, it also invariably raises new questions. Sometimes research or the emergence of new evidence casts doubt on what was previously regarded as solid knowledge, such as the value of irrigating the soils of arid regions, or the sustainability of logging old growth forests. Since it seems almost certain that advancing knowledge will reveal some current practices to be unsound, that advancement of knowledge should be a high priority. A small investment in research and development now may avoid irreparable damage later.

There is a more fundamental limitation on our ability to know the impacts of our actions on natural systems. Most of our modelling assumes we are making small, reversible changes to systems that are in equilibrium. The caution expressed by the IPCC applies more generally to non-linear systems:

> Future climate changes may involve 'surprises'. In particular, these arise from the non-linear nature of the climate system. When rapidly forced, non-linear systems are especially subject to unexpected behaviour.

This is an important warning. When we change the conditions applying to complex systems, we produce changes which will not be expected; some of these will be counterintuitive. We can now see some of the consequences of past actions; in some cases, we wonder why those consequences were not anticipated. It does not require detailed understanding of river systems to see that removing 99 per cent of normal water flow will produce significant changes to the riverine ecosystem, nor does it take much understanding of biodiversity to see that clear-felling of forests will put pressure on forest-dwelling species by destroying their habitat.

One recent piece of research illustrates the complexity of the interactions between species in natural systems. A study of truffles, the fruiting bodies of fungi, in the eucalyptus forests of south-eastern New South Wales showed the crucial role of the long-footed potoroo in the health of the overall ecosystem. The potoroo unearths and eats the truffles, then excretes the spores of the fungus, thereby making it available to other trees. The fungus becomes attached to roots of trees in a mutually beneficial symbiotic arrangement. So we now know that even a forestry economist who was interested only in the production of saw-logs should recognise the value of the long-footed potoroo to the health of the growing timber. This relationship has only been understood in the last few years.

There are undoubtedly many similar stories yet to be uncovered of the importance of apparently minor species to the health of ecological systems such as forests, grasslands or estuaries. What we now know, in general, is that the loss of any one species from a complex system will usually have flow-on consequences, and in some cases those effects will not have been predictable from our previous knowledge. So we need to invest more research effort into interdisciplinary studies of complex systems, integrating the disparate relevant fields of knowledge. As the Friibergh workshop mentioned earlier (page 47) concluded: 'we now urgently need a better general understanding of the complex dynamic interactions between society and nature. That will require major advances in our ability to assess such issues as the behaviour of complex self-organising systems, irreversible impacts of interacting stresses, various scales of organisation and social actors with different agendas.' Since some complex environmental problems have different possible explanations, we need to ensure that the research support process does not preclude study of the alternatives. Thus it is a high priority, if our policy framework is to be well informed, for our research system to be explicitly and practically pluralistic.

We also need to develop a process that recognises and values Indigenous ecological knowledge. Most decision making implicitly assumes that Western scientific knowledge is inevitably superior to Indigenous knowledge. Of course, there are many examples of scientific understanding underpinning modern use of natural resources, and there are many complex effects we now understand in ways that Indigenous Australians did not. This should not blind us to acknowledging that Indigenous people also have an understanding of remote parts of Australia that has allowed them to live and reproduce there, while non-Indigenous people regularly perish in those places or require multi-million dollar rescue operations. In some national parks, Indigenous understanding of natural systems is now used as part of the management process. This is a useful model for the future, valuing and incorporating relevant Indigenous knowledge. Just as scientific knowledge embodies theories and models which are at least as important as facts, so Indigenous knowledge incorporates metaphors and images which are also important. We need to acknowledge and respect those metaphors and images as well as the extensive factual knowledge about individual species or the location of water. This is an important dimension of the principle of inclusiveness, ensuring that the decision-making process recognises and values different knowings.

The need for environmental literacy makes obvious demands for changes to the process of education in general and science education in particular. I have argued that the traditional process of science education consists of 'the revelation of a fixed body of knowledge having almost divine authority'. This is not only alienating and as a result educationally questionable, but it also gives a totally misleading impression of the state of scientific knowledge, implying that it is a fixed body of eternal truths rather than work in progress. Science is not a stable body of knowledge but the *process* of trying to understand the natural world and our impacts on it. Teaching science as a body of knowledge would be like teaching politics without considering either the struggle to develop our democratic system from the age of divine right of the monarchy or the continuing discussions about such issues as our voting systems, ministerial responsibility, the role of the head of state and so on. The standard approach to science is actually more misleading, because most people working in the political system accept it as it is; only a few reformers are actively working to change the voting systems we use or to sever our outmoded links to the British royal family. By contrast, no serious scientist accepts our current knowledge as the last word. Every working scientist is actively striving to improve our understanding of the world by collecting more data, by improving theoretical frameworks or by challenging existing ideas. So environmental science has to be taught as a process rather than a body of knowledge, with explicit recognition of the levels of uncertainty in our current understanding, in terms both of basic knowledge of the local environment and general understanding of complex natural systems. It also needs to recognise that applying our understanding of the natural world to real decisions is inevitably a complex process that has social, political and economic dimensions. Decisions about the natural resources and environment need to use a longer time frame than has been usual in recent thinking, need an appropriate structure that allows an integrated or holistic approach, and need to recognise the primacy of ecological considerations rather than seeing them as an optional extra.

We should base our choices on a much longer time horizon than is the norm in political discourse. We need to use timescales of decades or centuries rather than weeks or months. The damage done to Australian natural systems by inappropriate practices has taken centuries to reach the point at which action is demanded, and will take decades or centuries to repair. This conclusion is not unique to Australia, but quite general. So the use of natural resources and the way we treat the environment need to be decoupled from the day-to-day adversary system of party politics, almost ensuring ad hoc decision making, and put on a secure long-term footing. This argument suggests that decisions should rest on the secure foundation of scientific knowledge. While this approach has an obvious appeal, its usefulness is limited by two fundamental problems. First, technical understanding is rarely clear and unambiguous. Secondly, even when our understanding is definite, values will always have a role in determining the response.

While we should try to ensure that research is not consciously biased, there is no prospect of research being independent of theories or underlying values. Our mental models or prevailing theories determine which data we collect, how we assess the results, which research programs we set up, which projects we fund and which researchers we see as credible. As Albury argues, the process of advancing knowledge is inevitably value-laden, as is its assessment. So there is always a social (and political) dimension to the decisions about which science is supported. Changes to the membership of research funding bodies can influence the balance of support between broad areas, while some ministers have been known to put a personal stamp on research programs. Deciding to ask a question does not guarantee that it will be answered, but we are less likely to find answers to questions we decline to ask. In an atmosphere of limited funding for research, a decision to fund one project always precludes the support of others.

Even when there is agreement to study a particular problem, if it is a complex issue there can be alternative explanations. If the data do not allow the issue to be resolved, different scientists will legitimately come to different conclusions, as in the debate about the source of the phosphorus in inland streams. The debate about whether there is a discernible human influence in the observed changes to the global climate is another example. While atmospheric scientists warned in 1985 that human actions seemed to be changing the climate, it took another decade for the wider scientific community to accept that conclusion and there are still some scientists who do not accept the dominant view. Again, values play a role in assessment of the data and the validity of models. Natural systems often do not allow controlled experiments, while the lack of baseline data and the long timescales involved make it unlikely we will achieve the goal of certain technical knowledge.

Finally, even if the technical knowledge is clear and unambiguous, the response always involves some balance between ecological needs and desired outcomes in other areas: economic, social or political. It has been understood for decades that the diversion of water from the Snowy River has affected its ecological values. Restoring even a small fraction of the previous flow of the Snowy has economic, social and political implications, so the issue has been on the political agenda for years. Similar comments could be made about the extraction of water from the Murray Darling river system, the use of the Great Artesian Basin, the stocking of some rangelands, clearing of vegetation from agricultural land and traditional irrigation practices, all issues of such complexity that they remain unresolved. Since differences about resource or environmental issues are usually based on differences in values, we need to recognise the role of values in the analysis of complex issues. Since there are legitimately (and inevitably) different acceptable values in a pluralistic democratic society, there will always be some degree of disagreement or conflict about environmental or resource issues. The process for resolving these disagreements needs to explicitly recognise and focus attention on the role of the underlying values.

This is simply an extension of a general argument advanced by the Ranger inquiry over 20 years ago, that the role of experts should be to provide technical information that allows the general public (or their elected representatives) to make decisions based on that information. The RAC made a similar case in their report on the proposal to mine Coronation Hill. The proposal was evaluated in economic terms, giving the range of estimates of the economic benefits to the mining company and to the broader community. It was evaluated in environmental terms, giving various assessments of the risk to the Kakadu wetlands of the proposed extraction process. It was also evaluated in social terms, assessing the possible costs and benefits for both the local Indigenous people and the broader community. Striking the balance between these considerations, deciding whether the economic benefits were worth the environmental risks and the social effects, was (as the RAC said) inevitably a value judgement. There is no prospect even in principle of some sort of modernised 'felicific calculus' which tells decision makers the 'right' choice to make. It is history that the Hawke Government decided not to approve the project, but many people disagreed with that decision; indeed, there is evidence that many members of Hawke's Cabinet did not agree with the prime minister, and the disagreement was one of the factors involved in Paul Keating's subsequent successful challenge to Hawke's leadership. So we must recognise that a secure technical understanding does not of itself ensure sound use of natural resources or environmental assets. In the final analysis, there will always be political factors influencing the way the technical understanding is used. So the development of environmental literacy has to include an appreciation of those political influences.

An important complication of decision making is a recognition of redistributive effects. Change always has costs as well as benefits, losers as well as winners. These effects should be explicitly recognised by accepting that there is some responsibility for those who benefit from policy changes to compensate those who lose. In the example of noise imposed on Sydney residents by the new second runway, air travellers now pay a hypothecated levy on top of the air fare, and the proceeds of the levy are used to compensate the residents affected by the resulting noise. While that system of losers being compensated by beneficiaries was a hasty response to an urgent political problem for the government of the day, it has established an important precedent. There is an obvious dimension of social justice in the principle that those who benefit from public policy should compensate those who are disadvantaged by that policy. Historically, there has been a tendency for policy formulation to reflect and reinforce existing power relations. It is difficult to justify that within a limited time frame, but it is impossible to defend when the inequity is intergenerational. Many of the choices made about natural resources or the environment have effectively disadvantaged all future generations for the benefit of the present generation. This is the moral equivalent of stealing from our own children. In those terms, it is morally indefensible.

The discussion of environmental literacy has obvious implications for education. It suggests that the content needs to be broader than an emphasis on 'environmental science', especially where that is interpreted narrowly as the body of knowledge about environmental systems. It has to include the social, economic and political dimensions of our interaction with natural systems. It must explicitly recognise that our engagement with natural systems is inevitably driven by social factors and steeped in the dominant values of the time. So it must be interdisciplinary in nature and based on a recognition of the complexity and uncertainty of the real world. Producing a credible solution for a complex problem is a much more demanding task than trotting out a well-rehearsed proof for an obscure theorem, listing the halogens by order of atomic weight or reciting the names of British monarchs. It requires real environmental literacy, including an appreciation of general principles, an awareness of the specific issues of the case being considered and a recognition of the limits of our current knowledge.

Not only is such an approach a better preparation for the making of wise decisions about the environment, it is also a generally better preparation for the complex world in which we now live. The old content-based education was effectively a preparation for a world of stability, in which a fixed body of knowledge and a defined set of skills was an adequate preparation. In the modern world of rapid change, where much of the knowledge and many of the skills that people will need in their future life do not yet exist and so cannot be taught, the formal education must emphasise the processes which will prepare people for that world. An explicit commitment to environmental literacy will therefore lead to a better educational preparation for the complex, rapidly changing world of the future.

EXPONENTS OF THE IMPOSSIBLE

The Weekend Australian, 30–31 March 1990

Many of our decision makers exhibit a tragic lack of numeracy. Politicians often have backgrounds in law, business, farming or trade unions. They are usually able to read balance sheets and interpret judicial decisions, but would not recognise a quadratic equation if they met one in the Members' Bar.

Many key decisions are taken by bureaucrats. The upper ranks of the public service are now dominated by a new freemasonry of market economists. With the rhetoric of market forces playing the role once filled by secret handshakes, they systematically promote their brothers into the positions of influence. While these economists are often oblivious to the social implications of their desire to establish an 18th century Utopia dominated by market forces, it might be expected that they would at least have mathematical insight. Alas, many seem unaware of the one piece of mathematical knowledge that all decision makers should have: the behaviour of the exponential function.

They cannot perceive the long-term implications of growth, usually seen as a desirable thing. Most of us are not so naive; we would be worried if our weight kept on increasing and puzzled if our height became greater each year. We know there is an optimum size for each of us. That personal realisation does not extend to organisations, which generally assume that continued growth is desirable. Despite all the evidence to the contrary, many decision makers seem to believe that large organisations are more efficient.

This belief has spurred the disastrous agglomeration of government departments to produce bloated bureaucracies which appear unmanageable. It also led to the bizarre amalgamation of efficient small universities and colleges into multi-campus organisations which spend large amounts of public money trying to manage their disparate components. Just as individuals have an optimum size, so do organisations. There are economies of scale up to a certain point, beyond which the problems of management reduce efficiency. The problems of the public service are the stock-in-trade of stand-up comedians, but the same comments could legitimately be made about large private corporations.

One of the most famous biologists of all time, JBS Haldane, wrote about the virtues of being the right size. For any species, he argued, there is an optimum size. Organisms that are too small are too weak to survive, those too large are too clumsy. The same argument applies to organisations. Those which are well managed aim at an optimum size for their style of work, a scale which brings maximum efficiency.

Many decision makers aim for growth, which is often described in terms of a percentage. Thus it is said that the economy is growing at 2 per cent a year, or that prices are increasing at 8 per cent a year. This type of growth is known as exponential growth: the expansion each year is a fixed percentage of the total. This means that the actual size of the increase becomes larger each year.

Suppose that the population of south-east Queensland is 1.5 million and it is growing at 3 per cent a year. The increase this year will be 3 per cent of 1.5 million, or 45,000 people, making the population 1.545 million. The increase next year will be 3 per cent of 1.545 million, or 46,350 people, making the total 1.591 million. The following year the growth will be 3 per cent of 1.591 million, or 47,740 people, taking the population to 1.639 million, and so on.

The most important characteristic of exponential growth is that it has a fixed time for each doubling in size. A rule of thumb is that the doubling time for any given rate of exponential growth is found by dividing the percentage rate into 72. Since 72 divided by 3 is 24, 3 per cent growth means a doubling every 24 years. Think of the implications of this. If a population of 1.5 million grows at 3 per cent a year, in 24 years it will be 3 million. After 48 years it would be 6 million, after 72 years 12 million and after 96 years, 24 million. Thus the apparently tolerable growth rate of 3 per cent would, within less than 100 years, expand the population of south-east Queensland from 1.5 million to an unimaginable 24 million.

This is the same principle behind the famous tale of putting one grain of rice on the first square of a chessboard, two on the next, four on the next and so on. Long before the last square is reached, the amount of rice needed exceeds the total production throughout all human history. Another illustration of exponential growth involves imagining a water lily that grows continuously, doubling its size each day so that it will cover the entire pond in 30 days. After 27 days, it only covers one-eighth of the pond, so there appears to be no real problem. After 28 days, it covers a quarter of the pond; you might begin to worry. After 29 it has covered half the pond and the situation is clearly serious. You then have only one day to prevent it completely covering the pond.

Accepting exponential growth means accepting a situation that will get worse at an accelerating rate. If the Gold Coast has trouble coping with present levels of sewage, as suggested by the recent studies of water quality, how will it cope with twice as much in 20 years? It is naive in the extreme to assume that the size of the human population, or the rate at which we exploit mineral resources or cut down our forests, can go on increasing in an exponential fashion.

If we do not recognise the implications of exponential growth, they will be drawn forcibly to our attention as natural systems fail to cope. If we recognise the problems, we can aim for a 'smooth landing' in a sustainable way of life. If we don't, a future minister for the environment may well be telling us about the ecological catastrophe we had to have. This is a strong argument for all politicians and senior bureaucrats being given a crash course in basic mathematics, emphasising the characteristics of exponential growth.

POSTSCRIPT: REFLECTIONS AFTER COPENHAGEN

'It is too early to say.' That was the reported answer of a 20th-century Chinese leader to a question about the impact of the French Revolution! As I finalised this book, it was certainly too early to say whether history will regard Copenhagen as another tragic missed opportunity, confirming the widespread view that our social institutions and political systems are incapable of responding to the challenges we face. There is still an outside chance that future historians may regard Copenhagen as a genuine transition which enabled civilisation to survive. But it really is too early to say.

Let me first summarise my experience of the conference. The 2009 UN climate change conference was held in Copenhagen over two weeks in December. I arrived on the Sunday in the middle of the meeting but, along with thousands of other people who arrived that weekend, was denied entry to the venue, the Bella Centre. I joined a long queue at 9.15 am Monday to go in and present my credentials and be issued with my pass. Media identities like Karen Middleton of SBS and Emily Rice of Channel Ten News were ahead of me in the very long line. It moved with glacial majesty until about 1 pm and then stopped completely. We were all still waiting outside, in freezing cold with flurries of snow, at about 4 pm when the UN official who was loosely in charge announced that they would give special attention to media representatives and invited them to form a separate line. Scrambling over metal barricades and passing cameras across, they massed into a separate scrum from the one for lesser mortals (scientists, non-government organisation people, etc). Then at 5 pm the same stand-up comedian announced that they would only accredit those already inside the tent. He invited us to return the next day at 8 am to take our chances in the second round of the lottery. The angry crowd started chanting 'Let us in!' and unprintable slogans about the UN officials. The little man retreated, leaving the puzzled Danish police to maintain order and keep us out.

We found out later that the venue was overbooked and there was a capacity problem. Apparently the UN had allowed over 30,000 people to register for a venue that holds 15,000, presumably hoping that most of them would not turn up in Denmark. When it became clear to the geniuses inside that we couldn't all be accommodated, the drawbridge was pulled up arbitrarily at about 1 pm Monday. I was in the line with two scientists who were expecting to present at events that afternoon and had flown to Copenhagen just for the day to do that; both were turned away (along with many media people). I found out later that government delegation leaders and such luminaries as the chairperson of the International Union for the Conservation of Nature were also left standing in the queue for hours. Others I met had flown from Japan on budget air tickets that allowed no change of dates, so they were stuck in Copenhagen with no access to the conference venue.

I made another attempt to get in on the Tuesday morning. We were told that the Monday problems were due to everyone – scientists, media, activists, pollies – all being in one queue, so we were sorted into separate queues and urged to be patient. After an hour, the queue for those representing NGOs had moved about 50 metres, with about 500 metres still to go even to reach the registration area. I did the sums and consulted colleagues by phone; they reported that email traffic was full of complaints about scientists and NGO people failing to gain admission. I admitted defeat and slunk off to a nearby hotel, where other would-be delegates were also holed up, monitoring proceedings on the web and sending text messages to colleagues inside the venue. As I said to one, if I was just going to do that I could have been by the beach at Marcoola, warmer and more comfortable as well as less jaundiced about the incompetence of the UN organisers! I couldn't help wondering whether the UN secretariat could organise the proverbial piss-up in a brewery, let alone a climate change conference, still less a coordinated global response to a major environmental problem.

There was better news on Wednesday! The Australian Conservation Foundation's multi-talented CEO Don Henry pulled some strings and got me accredited, after only about 40 minutes standing outside in the snow. He had earlier invited me into a breakfast meeting where Kevin Rudd explained to many of the Aussies in town what he hoped to achieve. Inside the venue was slightly less shambolic than outside. Only slightly, because the working groups had been grinding away on draft text until 6 am, only for the Danish prime minister to take over chairing the meeting at 9 am and announce that he would be producing a new text. Delegates from China and the developing world objected strenuously. The chair reminded them that the world was expecting the conference to make real progress to slow climate change, while the objectors complained that there was an issue of principle.

The impasse highlighted a serious problem. The United Nations consists of all its member states, about 190 of them. Clearly a meeting of one representative from each country is far too large and unwieldy to draft an agreement. That was shown by the slow progress in producing a document for the conference to consider. After two years of work on the so-called Bali Road Map agreed in 2007, the text was still a mess. A more effective process is to set up a working group representative of major identifiable different interests: large industrialised nations; smaller affluent countries; large developing nations; small island states; poor African countries; and so on. But when such a group is set up and produces a working draft, there are always some of those not included who protest that the process is not inclusive or transparent. That dilemma has bedevilled attempts to produce a legally binding instrument to respond to climate change. The large groups are painfully slow at the best of times; in this case, the negative influence of the fossil fuel lobby absolutely ensured that they made little progress. When UN officials or the conference chair tried to speed things up by convening smaller groups, those who didn't want to see effective progress were able to derail the agreement by objecting to their exclusion from the drafting process.

At the conference, large screens relayed to those of us in the outside areas a series of set-piece speeches. Every leader, even those with very dubious environmental credentials, made a speech claiming to be setting the standard and implying that the problem would be solved if only other nations were as good. The European Union called for ambitious targets and proposed reducing emissions 30 per cent by 2020 as part of a global agreement, while also suggesting ways of funding the energy transition in poorer countries. It also proposed halving deforestation rates by 2020 and stabilising net forest area by 2030. At the other end of the spectrum, Hugo Chavez of Venezuela blamed the capitalist economic system and said: 'If climate change was a bank, it would have already been saved!' While there was an element of theatre in his claim, it can't be denied that the cost of moving to a clean energy economy is less than has been spent bailing out financial institutions in 2008 and 2009. The leader of another small country argued that the world should consider establishing a climate tribunal, along the lines of the International Court of Justice in The Hague, where those who had knowingly inflicted climate damage on the world could be held to account. I could imagine how difficult it would be to recruit managers and directors to coal mining companies if that threat were ever implemented!

Having recently seen the extraordinary spectacle of the Federal Coalition being taken over by strange forces who deny the science of climate change, I was relieved to see that there was no such nonsense among the leaders in Copenhagen. More than 130 heads of state took the floor and urged concerted action to slow climate change. French President Sarkozy actually said, 'Who would come to this podium and say they are against concerted global action?' It was a rhetorical question. Every one of the leaders professed to support serious action, every one claimed to be already taking action in their own country.

The basic divide was clear. For the world's overall emissions of greenhouse gases to peak by 2015 and then trend steeply down, as the science says is necessary to bring the risk of dangerous climate change down to an acceptable level, there have to be serious reductions in the developed countries between now and 2020. For the downward trend to continue beyond 2020, the larger developing nations like China, India and Brazil must curb the expected growth in their greenhouse gas production. Each group says the other is not doing enough. Both are right. Despite the urgency of the situation and the desperate need for change, most leaders still behave as if the issue is a subsidiary one to be addressed only after economic considerations have been satisfied. While the leaders of Brazil, Russia and the US said their commitments were unconditional and would be fulfilled regardless of whether the rest of the world followed suit, those reductions are not enough to avoid the risk of serious climate change. The Australian Government's proposal of 15 per cent reduction, perhaps 25 per cent if improbable international conditions are met, is woefully inadequate in the light of the recent science.

There are other issues. There was trenchant criticism of attempts to expand the provisions of the so-called clean development mechanism to include unproven technology, like carbon capture and storage, as well as the proven dirty option of nuclear power. There was also a serious proposal from Europe to alter the provisions for counting land-use change from the Kyoto basis of a 1990 baseline. There was talk of a 'forward baseline', an amazing bit of creative accounting which would allow countries to claim credit for not destroying forests in the future. This ranks with the notorious 'Australia clause' which allowed the Howard Government to inflate our 1990 baseline by including the prodigious land clearing then happening in Queensland. My view is that a forward baseline only makes sense on a tennis court!

At side events during the conference, there were detailed technical presentations about the scale of change which could be achieved with political will. Studies of the UK, Denmark, California, Europe as a whole and even one Stanford University global study all came to the same conclusion. Those of us living in developed nations could live at the same level of material comfort using half as much energy as we now use. The technology exists to provide the same level of services – heating, lighting, mobility, electronic gadgetry – with half the present energy use. By 2030, all of that energy could be supplied by a mix of renewables. So there is no rational reason for avoiding strong action now.

It was heartening to meet the other Australian scientists and members of NGOs who are working away for a good outcome, struggling to keep our government honest as well as finding ways to influence the broader process. By the end of the conference, I was impressed by the energy and commitment of the NGO people in the building and close to despair about the lack of urgency shown by the politicians. We were all hoping for miracles to be achieved by the leaders when they arrived for the last few days of the conference. But even then, there was little sense of urgency. Posturing and point-scoring trumped promises of substance, with little sign of purposive action.

The last 24 hours bordered on farce. There was no prospect of the large groups providing text that would allow the leaders to make decisions. One analysis counted over 100 examples of 'bracketed' clauses, requiring a decision. ('In order to keep the temperature rise below [3/2/1.5] degrees, it will be necessary to reduce global emissions by [50/80/95] per cent by 2050.') In practice there were only four areas of substantial disagreement, as Kevin Rudd pointed out in his speech to the meeting, but some of them cropped up over and over again in the draft text. So US President Barack Obama, frustrated by the lack of progress, put together a group representing most of the main interests. It included the US, China, India, Brazil, South Africa and Ethiopia. They still had great difficulty reaching agreement across the traditional faultlines, but by the time Obama had to leave they had crafted a form of words that all could live with. He admitted that it did not go as far as some would have liked and did not go as far as the science demanded, but it was a concrete start. Then the 'Copenhagen Accord' was presented to the full gathering. Some nations predictably objected that the process had not been transparent and inclusive. Only a relatively small number objected to the text, but the development of a treaty works on the basis of consensus. So there was no way of approving the new form of words.

After hours of wrangling and long suspensions of the formal meeting to allow discussions offline, the conference eventually agreed (at about 10 am after meeting all night) to 'note' the Accord. That means it has no legal force at all. It was still an historical agreement. The Kyoto Protocol, adopted at the 1997 conference, only applied mandatory targets to the industrialised nations. It was weakened still further by the US refusing to ratify the treaty. So it did not restrict about two-thirds of the world's greenhouse gases, those emitted by the US, China, India, Brazil, South Africa, Indonesia and so on. The Copenhagen Accord was based on extending the Kyoto process to include new undertakings from the industralised world and broadening the agreement to include voluntary restraint by the large developing nations. It is the first genuinely global agreement to tackle the global problem.

Even though the delegates went home without any legal basis for the new arrangements, just a promise to work over the following 12 months leading up to the next conference in Mexico at the end of 2010, the Copenhagen climate change conference was a significant step forward. Over 100 world leaders agreed that urgent concerted action is needed to slow climate change. For the first time, all the major greenhouse gas emitters have agreed to be part of a global accord to tackle the problem. That includes the US, which never ratified the Kyoto Protocol, and the large developing countries that had no obligations under that agreement such as China, India, Brazil and South Africa. The Accord still needs to be turned into a treaty with legal force. There is a clear timetable for that to happen by this year's meeting in Mexico.

I was heartened at Copenhagen to find that everyone understood the importance of the issue. Nobody takes seriously the climate change denialists, a motley crew who don't even agree among themselves on anything except their starting point that we should do nothing. Every week the science is clearer and more alarming. It says we have only five to ten years to turn the upward emissions trajectory into a downward trend. The arguments are about which nations or industries will do what, and who will pay for the transition.

Now the developed nations – including Australia – have to put forward serious plans for the scale of emissions cuts needed, toward 40 per cent by 2020. As US President Obama said at the last night of the conference, the targets being put forward today are not sufficient. The science demands more aggressive action. Kevin Rudd has to stand up to the big polluters and set serious emission reduction targets. A package of measures to effect real change in Australia must go well beyond the proposals in the watered-down Carbon Pollution Reduction Scheme (CPRS), rejected by the Senate in late 2009. That was still too demanding for the Opposition and brought the denial faction to power in the Coalition, so the government has a real political challenge. But the conditions set by the Rudd Government for going to a 25 per cent target have largely been met. The science says we should go further. There is no economic or social reason to delay.

As well as reduction commitments from the industrialised world and corresponding commitments to meaningful action by the large developing nations, the Copenhagen Accord puts money on the table to help poorer countries adapt to climate change and manage the transition to clean energy. Pollution reductions must be measurable, reportable and verifiable. Everyone agrees with the principle. The challenge is finding mechanisms that command respect without infringing national sovereignty. The problems of nuclear technology remind us how difficult those requirements are.

Opinion polls show that the community understands the issue and wants to see concerted action. Tens of thousands of Australians joined the Walk Against Warming on 12 December 2009. Huge numbers have insulated their houses, bought solar panels and shifted to public transport. Now it is time for our leaders to follow the lead of the people.

Ten years ago, at the end of the 20th century, I was asked to answer the question, 'Can our civilisation survive the 21st century?' I said at the time that there was a chance, but if you were a rational gambler you would want very generous odds to bet on it. Most of our leaders still show no sign of even recognising the scale of the problems we face, let alone having the will or the political skill to build support for the sort of changes needed. The Copenhagen conference reinforced that gloomy view of our future. On the other hand, after I recovered from the Copenhagen experience I made my annual pilgrimage to the Woodford Folk Festival. As well as enjoying the rich diversity of music, I was involved in seven sessions of discussion about climate change, broader issues of sustainability and more specific issues like population. I was tremendously heartened by the extent to which people understand the nature of our predicament. There are really now only two paths forward, one leading to the gradual disintegration of the essential services for a civilised future and the hopeful alternative of a fundamental change. When the people lead, the leaders have to follow. It reminded me that the future is not somewhere we are going, but something we are all creating. Which one eventuates is up to us. If you have read this far, I hope you have been encouraged to join those of us who are working actively for a sustainable future.

ACKNOWLEDGEMENTS

I gratefully acknowledge all those who have encouraged me to write, especially teachers at Caragabal State School, Tahmoor State School and Bowral High School, and editors who have published my writings, especially Michael Haynes as editor of *Tharunka* in 1963, David Armstrong who edited *The Australian* in less ideologically blinkered times, the late Ian Anderson and Dr Rachel Nowak as Australasian editors of *New Scientist*, Paul Clifton as editor of the *nteu advocate* and Guy Nolch as editor of *Australasian Science*. I thank Black Inc, Allen & Unwin, CSIRO Publishing, University of New South Wales Press and Scribe Books for permission to reproduce material they originally published. I thank my academic colleagues and generations of students for batting around the ideas that have resulted in the writings collected in this book. I will be eternally thankful to my mother for ensuring that I continued my education. Dr Patricia Kelly has not only been my partner and creative sounding board, but has also immeasurably improved my writing with her insightful critique and attention to clarity. Finally, it has been a real pleasure to work with Alexandra Payne and her colleagues at University of Queensland Press, who were so helpful in producing this book.